Global Climate and Energy Governance

Tracing the changing activities of international bureaucracies active in global climate and energy governance, this book provides an in-depth analysis of processes of institutional innovation and governance integration between the two fields.

It shows that rather than the consequence of a designed strategy, governance integration – the convergence of approaches and practices among different actors within one or between two or more governance architectures – has come as the result of organizational changes arising from the international bureaucracies' various efforts to pursue and broaden their mandate in a complex and dynamic global policy environment. Each of the three cases analyzed (the UNFCCC Secretariat, the IEA Secretariat, and the World Bank) began their life focused on particular activities that today, following periods of sustained organizational change, make up only part of their operations. Beyond creating greater synergies for cooperation across the governance architectures, improving policies, and mobilizing greater investment to tackle the climate emergency, the book shows governance integration to have contributed to preserving and expanding the role and relevance of all three international bureaucracies.

This book will be of interest to students and scholars of global climate and energy governance, climate policy, and international organizations and their bureaucratic arms. Practitioners will find this book useful in thinking about why innovation in governance emerges and how it may be directed.

Harald L. Heubaum is Senior Lecturer (Associate Professor) in Global Energy and Climate Policy at the Centre for International Studies and Diplomacy, Department of Politics and International Studies, SOAS University of London. He is also Deputy Director and Co-Founder of the SOAS Centre for Sustainable Finance.

Global Institutions

Edited by Thomas G. Weiss, *The CUNY Graduate Center, New York, USA* and Rorden Wilkinson, *University of New South Wales, Sydney, Australia*

About the series

The "Global Institutions Series" provides cutting-edge books about many aspects of what we know as "global governance." It emerges from our shared frustrations with the state of available knowledge—electronic and print-wise—for research and teaching. The series is designed as a resource for those interested in exploring issues of international organization and global governance. And since the first volumes appeared in 2005, we have taken significant strides toward filling many conceptual gaps.

The books in the series also provide a segue to the foundation volume that offers the most comprehensive textbook treatment available dealing with all the major issues, approaches, institutions, and actors in contemporary global governance. The second edition of our edited work *International Organization and Global Governance* (2018) contains essays by many of the authors in the series.

Understanding global governance—past, present, and future—is far from a finished journey. The books in this series nonetheless represent significant steps toward a better way of conceiving contemporary problems and issues as well as, hopefully, doing something to improve world order. We value the feedback from our readers and their role in helping shape the on-going development of the series.

A complete list of titles can be viewed online here: https://www.routledge.com/Global-Institutions/book-series/GI.

Small Countries, Big Diplomacy
Laos in the UN, ASEAN and MRC
Alounkeo Kittikhoun and Anoulak Kittikhoun

Global Climate and Energy Governance
Towards an Integrated Architecture
Harald L. Heubaum

Global Climate and Energy Governance

Towards an Integrated Architecture

Harald L. Heubaum

Routledge
Taylor & Francis Group

LONDON AND NEW YORK

First published 2022
by Routledge
2 Park Square, Milton Park, Abingdon, Oxon OX14 4RN

and by Routledge
605 Third Avenue, New York, NY 10158

Routledge is an imprint of the Taylor & Francis Group, an informa business

© 2022 Harald L. Heubaum

The right of Harald L. Heubaum to be identified as author of this work
has been asserted in accordance with sections 77 and 78 of the Copyright,
Designs and Patents Act 1988.

Trademark notice: Product or corporate names may be trademarks or registered trademarks,
and are used only for identification and explanation without intent to infringe.

British Library Cataloguing-in-Publication Data
A catalogue record for this book is available from the British Library

Library of Congress Cataloging-in-Publication Data
A catalog record has been requested for this book

ISBN: 978-1-138-95815-9 (hbk)
ISBN: 978-1-03-216358-1 (pbk)
ISBN: 978-1-315-66133-9 (ebk)

DOI: 10.4324/9781315661339

Typeset in Times New Roman
by Newgen Publishing UK

Contents

Acknowledgments

No creative effort is ever truly the work of one individual alone. This book, too, would not have been possible without a community of support to whom I am truly grateful. In particular, I wish to thank the many current and former officials at the UNFCCC Secretariat, IEA Secretariat, and World Bank who so freely gave of their time and knowledge in interviews and background conversations, helping to shed light on processes of transformation and change in each of the three international bureaucracies. Organizational change and, ultimately, governance integration require agency and action. Together with others, these officials continue to work tirelessly to find solutions to some of the biggest challenges of our time and make the world a better place.

Abbreviations

ADB	Asian Development Bank
AfDB	African Development Bank
AR	Assessment Report
CBDR	common but differentiated responsibilities
CCS	carbon capture and storage
CCXG	Climate Change Expert Group
CDM	Clean Development Mechanism
CFU	Carbon Finance Unit
CIF	Climate Investment Fund
COP	Conference of the Parties
CO2	carbon dioxide
CTF	Clean Technology Fund
CTI	Climate Technology Initiative
EIB	European Investment Bank
EU	European Union
EU ETS	European Union Emissions Trading System
FSB	Financial Stability Board
GCC	Global Climate Coalition
GCF	Green Climate Fund
GEF	Global Environment Facility
G8	Group of Eight
G20	Group of Twenty
GHG	greenhouse gas
GW	gigawatt
IBRD	International Bank for Reconstruction and Development
IDB	Inter-American Development Bank
IEA	International Energy Agency
IFI	international financial institution
IMF	International Monetary Fund
IO	international organization

IPCC	Intergovernmental Panel on Climate Change
IR	international relations
IRENA	International Renewable Energy Agency
JI	Joint Implementation
MDB	multilateral development bank
NAZCA	Non-state Actor Zone for Climate Action
NDC	nationally determined contribution
NGO	non-governmental organization
OECD	Organisation for Economic Co-operation and Development
OPEC	Organization of the Petroleum Exporting Countries
PCF	Prototype Carbon Fund
PV	photovoltaic
SCF	Strategic Climate Fund
SDGs	Sustainable Development Goals
SEforALL	Sustainable Energy for All
SFDCC	Strategic Framework for Development and Climate Change
TCFD	Task Force on Climate-related Financial Disclosures
UN	United Nations
UNCED	United Nations Conference on Environment and Development
UNDP	United Nations Development Programme
UNEP	United Nations Environment Programme
UNFCCC	United Nations Framework Convention on Climate Change
UK	United Kingdom
US	United States
WBG	World Bank Group
WEO	World Energy Outlook
WMO	World Meteorological Organization

Introduction

The case for governance integration

This book is about organizational change and how this change, observed in this study through the prism of three international bureaucracies active in the climate and energy fields, can lead to the greater integration of heretofore largely separate governance architectures. Processes of governance integration, understood as the convergence of approaches and activities among different actors within one or between two or more governance architectures, are occurring in a complex and dynamic global policy environment with no shortage of risks and challenges. Ending poverty and hunger, providing clean drinking water for all, halting the spread of dangerous communicable diseases and pandemics, ending conflict and violence wherever it may occur, responding to climate change, achieving gender equality – the sheer scale of these and a raft of other planetary objectives incorporated in the United Nations Sustainable Development Goals (SDGs) – put any meaningful progress beyond the reach of any one country or international organization (IO), instead requiring meaningful collective action. Notwithstanding a lively debate on the usefulness of having a large number of goals and targets, the SDGs now directly engage with critical issues the preceding Millennium Development Goals (MDGs) did not address, in effect shifting their overall focus. As Browne and Weiss observe, the majority of the SDGs are now either "wholly or partially concerned with managing resources, energy or climate change" given the strong environmental connotations of "sustainability."[1] However, beyond mere semantics, the emphasis on climate change is logical and consequential, as success in addressing many of the development goals, targets, and indicators is directly dependent on whether or not the growth in global greenhouse gas (GHG) emissions can be reined in commensurate with a rise in global average surface temperatures of "well below 2°C above pre-industrial levels" as prescribed by the Paris Agreement on Climate Change.[2]

DOI: 10.4324/9781315661339-1

Successfully mitigating climate change requires dramatic changes in the ways humanity produces and consumes energy. Since the early days of the Second Industrial Revolution in the mid-nineteenth century, carbon dioxide (CO_2) emissions from fossil fuel combustion have increased at an unprecedented speed and scale.[3] Emissions grew from 198 Megatons globally in 1850 to 33.3 Gigatons globally in 2019, a 168-fold increase.[4] Due to their CO_2 emissions profile, energy production and consumption – by businesses, industry, and private households – are the most important drivers of anthropogenic, that is human-made, climate change by far. Roughly two-thirds of all GHGs emitted today are directly connected to energy, and more specifically fossil fuel combustion, requiring a balancing of the need for a swift decarbonization of the energy system to meet climate change mitigation targets with ever greater demands for affordable energy to both sustain global economic growth and combat energy poverty.[5]

A strong, sustained growth in renewable energy capacity has been at the heart of the transition to the low-carbon energy systems of the future. And this growth has accelerated in recent years. According to the International Renewable Energy Agency (IRENA), 176 GW of renewable electricity generation was added to the global mix in 2019, accounting for more than two-thirds of all the new power installations that year and outpacing all the fossil fuel capacity additions combined.[6] And yet despite these changes, the global dependence on fossil fuels for electricity generation, heating and cooling, transportation, and industrial processes has continued. In addition, GHG emissions continue to be generated through the consumption of energy related to land use, notably in the clearing of land, the use of farming machines, oil-based plant fertilizer, and biofuel agriculture. The Intergovernmental Panel on Climate Change's (IPCC) 2018 special report on global warming of 1.5°C estimates that if the current GHG emissions trajectories continue, it will be impossible to keep the rise in global average surface temperatures to below 2°C above pre-industrial levels.[7] Shooting past this target would, in turn, have dramatic implications for soils, plants, water resources, animal and human life, and livelihoods across the Global South and North.

The rationale

It is clear that climate change and energy need to be addressed in tandem if the world is to have a real chance at avoiding the high risks and potentially catastrophic consequences associated with moving well past 1.5°C and heading toward 3°C or 4°C of warming above pre-industrial levels. Given the long-recognized interlinkages between global energy production and consumption on the one hand and climate change on the other, it is quite surprising that there has historically been only limited policy interaction, let alone integration, between the two governance fields. For years, the climate convention process did not directly define the climate change problem as one largely about energy use. The 1997 Kyoto Protocol mentions energy only six times (of which twice in the Annex); there is no single mention of fossil fuels or coal, and oil and gas are only mentioned once in Annex A.[8] Many country delegations to official negotiations under the United Nations Framework Convention on Climate Change (UNFCCC) have historically been led by environment and foreign ministry representatives rather than officials from ministries responsible for energy or natural resources, underlining the policymaking disconnect between energy systems and their environmental impact.[9]

The same problem manifests at the national level. Although since the late 2000s some countries have sought to integrate responsibility for energy and climate policy by creating new ministries – prominent examples being Denmark's Ministry of Climate and Energy (created in 2007) and the UK's Department of Energy and Climate Change (created in 2008 and then dissolved again in 2016, its parts integrated into a bigger business ministry) – institutional separation remains the norm, and compartmentalization has been observed to continue even after location of national policymaking responsibility for both fields in the same ministry.[10] Moreover, while the majority of the world's governments have passed climate policies and set national GHG emissions targets for the medium and long term, only few have also managed to design, let alone successfully implement, energy policies in line with the goals of the Paris Agreement and the acknowledged urgent need to swiftly decarbonize their economies.

With a patchwork of activities at national level, significant moves toward the integration of energy and climate policy have emerged in unexpected ways at the international level. This book is about the developments that drive and have made possible this integration process: organizational changes in international bureaucracies. It does so by tracing and analyzing the changing activities of three key international bureaucracies: the UNFCCC Secretariat, the International Energy Agency (IEA) Secretariat, and the World Bank. It shows that, due to a number of internal and external drivers, each has become an important and influential agent in the integration of governance architectures. The research fills a demonstrable gap in the literature on global climate and energy governance as discussed below. In its analysis of the drivers of organizational change, it also speaks directly to the literature on international bureaucracies. This research is driven by a greater need to disaggregate the organizational black box and trace routine connections within and beyond. The international bureaucracy literature is able to capture dynamics that have been neglected by conventional approaches in which organizations are often just descriptors uncritically used, rather than an analytical category themselves.

The problem

Where can we turn to make sense of governance integration? The literatures on climate and energy governance have offered little in the way of an answer to date, owing to their evolution as if in two separate streams and for a long time mirroring the separation between energy and climate change policy and governance architectures addressed above. While the International Relations (IR) literature addressing the management of global environmental issues dates back to the 1972 United Nations Conference on the Human Environment in Stockholm, the field did not fully "come into its own" until the late 1980s and 1990s.[11] It is during this time that IR scholars began to seriously address the formation of international environmental regimes and institutions.[12] During the following two decades, the field grew further while also incorporating the new concept of global governance as different from an international, state-centered perspective. Scholars began conducting more research into the role played by IOs, non-governmental organizations (NGOs), transnational advocacy networks, and business actors.[13] Global climate change governance (as a particular kind of environmental governance) has in recent years received significant scholarly attention, with a wide range of publications focusing on all aspects of the global climate governance architecture and the UNFCCC Secretariat as the administrative entity at the heart of this architecture.[14] An important strand of this literature has been an analysis of the fragmentation and polycentricity of the global climate governance architecture.[15]

Influenced by the 1970s oil crises, much of the earlier academic research on energy policy focused on questions of energy security and developments in world oil markets.[16] Research on global energy governance, defined here as the "international collective action efforts undertaken to manage and distribute energy resources and provide energy services,"[17] has only emerged more recently.[18] Subfields to this literature have endeavored to analyze the role of specific organizations in a more complex world in which energy policy decisions are shaped by a multitude of actors across the Global North and South, including the IEA,[19] the G20,[20] the Organization of the Petroleum Exporting Countries (OPEC),[21] and IRENA.[22]

Like its counterpart in global climate governance, the literature on global energy governance has identified the fragmentation of governance architectures as a key challenge.[23] Even more so than global climate governance, global energy governance consists of a large number of actors that are not fully interlinked or integrated. Although there is no core organization that unites all countries under a single roof, the IEA is often seen as occupying a key role in global energy governance as "the world's gold standard for energy analysis"[24] and "the single most important institution for energy importing countries."[25]

However, while fragmentation within global energy governance and within global climate governance are increasingly well-researched and understood, there are only few publications today that have attempted to address existing gaps between the two architectures, taking a joined-up look at the two fields which does not simply treat one as the addendum of the other.[26] This includes attempts to understand planet-wide transformation through the more recent, holistic paradigm of earth system governance that incorporates various different governance agendas at all levels of policymaking, including environment, energy, food, and water.[27] What is missing beyond these accounts, however, is more detailed attention to the alignment and convergence under way between the climate and energy governance architectures and the organizational changes inherent to these. This is problematic given the closely intertwined nature of climate change and energy discussed above and the need to better understand how integration can benefit both. This book is therefore a contribution to addressing this neglect.

Explaining change

If governance integration has not received much scholarly attention to date, how are we to make sense of its causes? How can we explain the changes observed to have taken place in international bureaucracies in recent years that, in turn, form the necessary precondition for governance integration to occur? This study builds on the discussion of organizational change and innovation in the nascent literature on international bureaucracies that is in itself closely related to IR and its long-standing engagement with IOs. It adds to this some aspects of the public policy literature insofar as they help shed more light on the specific change dynamics encountered in the three cases analyzed herein. The study returns to both extant literatures in the in-depth discussion of the conceptual and analytical framework in 1. For now, the main relevant outputs are summarized.

The objects of this study are the bureaucratic arms of IOs (IEA Secretariat, World Bank) or convention/treaty processes (UNFCCC Secretariat), rather than the states that created them. Despite some more recent additions to the study of international bureaucracies, the wider IR literature has to date tended to emphasize analyses of the role of the latter in shaping IOs and regime-building over a deeper, more systematic engagement with the former, making the study of international bureaucracies a "fringe issue in the social sciences."[28] Over the years, more traditional explanations of changes in IOs as the result of changing state preferences have given way to accounts emphasizing internal organizational dynamics. A growing number of studies have produced insights into the role, autonomy, influence, and effectiveness of IOs, and the international bureaucracies that serve them.

The degree of autonomy afforded to an international bureaucracy by the IOs' member states is important in determining the bureaucracies' actions and choices when faced with new challenges. Much of the research on IOs and their bureaucratic arms since the 1970s has focused on this autonomy and what they make of it.[29] Given the observed changes in each of the three cases discussed herein, it is clear that a greater degree of autonomy and independence enables changes in activities to be driven in unexpected ways: from within the bureaucracy rather than through influence or pressure from states. The UNFCCC Secretariat, the IEA Secretariat, and the World Bank all operate with (varying) degrees of autonomy from the states that created them and/or constitute their membership, making them useful case studies in gaining an understanding of organizational change and governance integration. And yet autonomy alone is an insufficient explanation for change. International bureaucrats may well find themselves in an environment affording them greater freedom to make independent decisions but this tells us little about whether or not they choose to do so, which actors in the bureaucracy pursue which course of action, to which challenges they might react, and precisely how they would go about introducing new and innovative approaches to the activities of the international bureaucracy.

In another strand of the literature, building on studies of organizational autonomy, scholars have sought to explain the effectiveness and influence of international bureaucracies as more or less independent actors on the world stage.[30] This study does not intend to assess the influence or effectiveness of the UNFCCC Secretariat, IEA Secretariat, and World Bank. It rather takes as a given that each is an influential player in the realms of energy and climate governance and policy. And yet this literature has nonetheless produced further useful insights into the internal dynamics of international bureaucracies, offering a better understanding of the importance of their organizational setup, structure, and leadership. A number of studies have pointed to entrepreneurial leadership as key to improved organizational performance and greater influence of international bureaucracies,[31] in a similar way to the attention paid by public policy scholars to the key role of policy entrepreneurs in driving policy change.[32] A closer look at the public policy literature's analysis of the role of policy entrepreneurs is particularly apt given both its detailed theoretical development and its depth of empirical evidence produced to date. And yet organizational setup, structure, and leadership also only address some, although certainly important, parts of the bigger picture.

More holistic accounts that systematically trace and make sense of changes in the behavior and activities of international bureaucracies remain few and far between. Haas's approach proposes three models to understand change in the activities of, and decision-making within, international bureaucracies.[33] Built around organizational adaptation to a changing global policy environment and organizational learning, his study aims to show how changing problem definitions lead to behavioral changes. Here, too, the similarities to accounts of policy change within the public policy literature are striking, where learning from and adaptation to external developments are understood as key features of change mechanisms.[34] Haas's separation into "incremental growth" and "turbulent nongrowth" to characterize the types and degrees of organizational change observed in international bureaucracies is also mirrored by the public policy literature's distinction between incremental change and radical change punctuating longer periods of policy stability.[35]

An important aspect emphasized in the public policy literature but largely ignored by international bureaucracy scholars is the role of focusing events in helping to open windows of opportunity for policy change to occur.[36] These provide policy entrepreneurs with a temporary opportunity to alter course or establish new activities in ways that may not otherwise have been possible. For example, for the purposes of this study, negotiations to and passage of the Kyoto Protocol, the Paris Agreement on Climate Change or Assessment Reports published by the IPCC may all be seen as focusing events that, taken together, build an ever-growing case for climate action and create opportunities for change advocacy. Taken together, the variety of factors studied in both the IR-focused literature on international bureaucracies and the public policy literature can provide a fruitful approach to understanding changes in the activities of international bureaucracies that, in turn, act as the main drivers of governance integration.

The argument

This study proposes an explanation of processes of governance integration that fills the gap in the literature on global climate and energy governance identified above and adds to the literatures on international bureaucracies and public policy by tracing change within those bureaucracies. While the explanation falls short of constituting a theory, the arguments and reasons spelled out below offer a conclusive account of the changes observed to have taken place in the UNFCCC Secretariat, the IEA Secretariat, and the World Bank.

The study's first central argument is that integration of policy and governance architectures between the issue areas of climate change and energy has been under way for some time despite concurrent processes of governance fragmentation. Governance integration – that is, the convergence of approaches and practices amongst different actors either within one or between two or more governance architectures – is shown to be on-going in three major international bureaucracies each of which occupies a key role in the politics of energy and climate change.

Further, governance integration is argued to occur as the consequence of organizational changes within international bureaucracies that lead them to change or adjust their approaches and activities in ways establishing interlinkages with the approaches and activities of another actor or group of actors. In the context of this study, it would require an international bureaucracy operating in global energy or climate governance or, indeed, in a related field, to adapt its views and approaches and start behaving in ways that align its activities with that of another international bureaucracy, operating either within global energy governance or within global climate governance. This organizational change is broadly defined as the change in the commitment of an organization (here an international bureaucracy) to an objective, resulting in the adoption of new approaches and activities. It is these activities that this study sets out to trace and analyze in the three case studies presented below.

The study finds that the three international bureaucracies' increasingly consequential role in integration between the climate and energy governance fields derives not from a designed strategy or top-down plan, but emerges through their various efforts to pursue and broaden their mandate in a complex and fast-changing global policy environment. That is, rather than simply as a response to explicit demands from its member states, the three bureaucracies' role in integrating global climate and energy governance emerges through organizational change and adaptation driven by external challenges and novel ways in which each is exercising its organizational autonomy. Thus, the study offers a causal change mechanism adapted from international bureaucracy and public policy studies. It argues that the observed changes occur as a consequence of the interplay of at least four factors:

- A changing global policy environment that challenges international bureaucracies to adapt and provides windows of opportunity to do so,
- The degree of autonomy granted to the bureaucracy by the states that created it in the first place that actors inside the bureaucracy may draw on to set out a different direction,
- Structural changes such as the creation and growth of units and the broadening of expertise within the bureaucracy that introduce new thinking and new understandings, and, critically,
- The leadership of policy entrepreneurs who by virtue of their elevated position can shape the direction of the entire bureaucracy and its external positioning.

These entrepreneurs are critical in defining and re-defining problems and advocating potential solutions. Thus, organizational change and, subsequently, governance integration are the result of a complex interaction of a number of external and internal aspects.

Process-tracing within case studies

The analysis explores the role of three key international bureaucracies in the climate and energy governance space and discusses the ways in which each has changed its approaches and activities. It does so through a detailed tracing of processes in each bureaucracy. Detailed qualitative process-tracing does not normally allow for large-N analyses. Small-N analyses on the other hand are not normally broadly generalizable across cases. This study strives for contingent generalizations that "present limited risks of extending these conditions to causally dissimilar cases."[37] While all the international bureaucracies (and others beyond those studied here) find themselves embedded in the same changing global policy environment, internal factors, such as the main players in the international bureaucracy acting as policy entrepreneurs, are different and unique to each case. Therefore, results derived from the cases studied in this work may not be directly transferable, although the approach that underpins the research may well be applicable to a wide range of other international bureaucracies or, indeed, other state and non-state actors.[38]

Analyzing case studies is one of the most widely used methods of social science research today, delivering empirical data and contributing to theory development. According to Swanborn, they are "the study of a phenomenon or a process as it develops within one case."[39] Case studies allow researchers "to achieve high levels of conceptual validity, or to identify and measure the indicators that best represent the theoretical concepts the researcher intends to measure."[40] Organizational change as the precondition for governance integration is a multifaceted process, with a wide range of different causal effects and actors involved, leading to a variety of interaction effects. A detailed analysis of a small number of cases allows for the tracing of these effects better than statistical studies which "necessarily leave out many contextual and intervening variables."[41] Complex causality, particularly the problem of equifinality (an outcome can be the result of many potential causal paths), is ever present. The study has, therefore, turned to process-tracing to map out a causal path consistent with the argument and the evidence in each of the three international bureaucracies.

Being able to capture causal mechanisms in action within a case depends on the close examination of intermediate steps taken to arrive at the outcome of interest.[42] Telling a causal story for each of the three international bureaucracies discussed in this study is, then, dependent on integrating the views of officials on the inside as well as external experts closely connected to or with detailed knowledge of the activities of the UNFCCC Secretariat, the IEA Secretariat, and the World Bank, respectively. It is they who either closely observed the changes or helped bring them about in the first place. A number of semi-structured, open-ended interviews were conducted, both in person and over the phone. In total, 22 senior officials and a further ten external experts, of whom several had had prior connections to one of the three bureaucracies, were interviewed. The choice of interviewees was reliant on whether officials were located within sections of their respective bureaucracies addressing energy and/or climate change and on whether they were in a position to authoritatively address changes in the organization's activities over time. In addition to formal interviews, a number of background conversations helped shape the thinking on organizational change in international bureaucracies and the implications for governance integration.

As most of the interviews were granted on the condition of anonymity, they have been referred to in a non-attributable way throughout.

The names mentioned in the book are those of senior leadership personnel in the international bureaucracies that have appeared in official publications, news articles, and academic writing. For each interview drawn on for direct quotes, a general reference to the international bureaucracy has been provided, in addition to the date on which the interview was conducted.

In parallel to conducting interviews, a large number of official publications, speeches, website information, and news reports relevant to climate and energy policy and the interlinkages between them for each of the international bureaucracies were examined. Official publications included shorter reports, more in-depth documents, guidelines, or strategy papers published by each organization, for example, those spelling out the organization's direction of travel on climate policy, setting out recommendations for renewable energy or energy efficiency policies, or energy sector lending guidelines. Speeches were mainly those delivered by the organizations' top brass: the IEA's Executive Directors and Chief Economists, the UNFCCC's Executive Secretaries, and the World Bank Group's Presidents and other senior officials. International news reports were chosen for their relevance to the analysis, representing a wide variety of authoritative sources.

Why this matters

The answers this study seeks to provide are timely and relevant for several reasons. First, while there is now a detailed understanding of the processes and implications of governance fragmentation, especially in the environmental and climate governance fields, the mechanisms and implications of governance integration have been researched less and, consequently, remain less well understood. The first core contribution of this study is, thus, the analysis of emergent integration between the climate and energy governance fields and the effects of such integration, something largely neglected thus far in the extant governance literature, especially given the long-standing separation between the fields highlighted earlier. By building on Heubaum and Biermann in its delimitation of governance integration as a discrete concept, this study also offers a definition of a term that has so far been largely absent from the scholarly debate.[43]

The study's second core contribution is to analyses of the changing role of international bureaucracies. A detailed analysis of the changes in approaches and activities of the three international bureaucracies discussed herein – the UNFCCC Secretariat, the IEA Secretariat, and the World Bank – adds depth to an IR literature that has neglected the study of international bureaucracies for a long time and in the extent to which it addresses them has focused more on their autonomy, influence, and effectiveness. Organizational change and the various factors that explain it have been conceptualized by drawing on both the literature analyzing international bureaucracies and the public policy literature. The resulting change mechanism explains the observed developments in detail, giving credence to the argument that rather than as the result of a designed strategy or top-down plan enforced by member states, governance integration emerges through the international bureaucracies' various efforts at pursuing and broadening their mandate in a complex and dynamic global policy environment.

In its analysis of organizational change in international bureaucracies, the study also adds to the public policy literature which for many years has focused on analyses of policy change and the forces driving this change. In each of the cases presented here, a number of policy entrepreneurs acted as change agents, aided by changes in a global policy environment more generally. Drawing on insights from public policy adds depth to existing IR-driven explanations. Importantly, the study adds an international dimension to a public policy field that is still largely shaped by analyses of cases rooted within domestic political contexts.

Fourth, a comparison of organizational change and integration processes is illustrative because all three international bureaucracies serve international bodies occupying different spaces within global governance architectures: the UNFCCC as an international treaty process, the IEA as an intergovernmental organization within the framework of the OECD, and the World Bank as an international financial institution (IFI) focused on the Global South.

Finally, this study also points to governance integration as an important factor in improving climate and energy governance and policy. Climate change is now widely understood as one of the most important, complex, and urgent global problems of the twenty-first century, cutting across all aspects of economic and social life in the Global North and South, affecting actors on all levels of human organization, and requiring ambitious policies and financial flows consistent with the goals of the Paris Agreement. IOs and their bureaucracies operating in the climate and energy fields have a varied role to play, from the provision of data and analysis, policy advice, global advocacy, financial support, and catalyzation of private sector capital, to bridging the disconnect that has characterized the relationship between climate, energy, and related sectors for too long. The climate emergency cannot be effectively addressed without a focus on energy production and consumption at the heart of the problem. Likewise, the global energy system is in a period of deep transformation due to the pressing need to avoid the grave risks of unmitigated climate change and achieve a climate-neutral economy with net-zero CO_2 emissions by the middle of the century. IOs and their bureaucracies in the climate and energy fields would be abdicating their duty if they did not address these interconnections head on and worked to prove that given the right handling, no international challenge, even one as complex and intractable as anthropogenic climate change, may be insurmountable. By integrating and aligning their approaches and activities more closely, the international bureaucracies analyzed herein create greater synergies within and between the governance architectures, enhancing the effectiveness of both climate and energy governance and contributing to achieving the goals of the Paris Agreement and, relatedly, the SDGs.

Plan of the book

The study unfolds in six chapters. Chapter 1 spells out the analytical framework in greater depth, beginning with an understanding of governance integration as a discrete concept and continuing with an elaboration of organizational change and the international bureaucracies in which this change has been observed to have taken place. It then identifies the changing approaches and activities that are evident from the three cases, followed by a conceptualization of the four factors, or variables, constituting the change mechanism proposed herein: global policy environment, organizational autonomy, organizational structure, and organizational leadership. Although there is no agreed upon definition in the extant literature, the first variable, the global policy environment, precedes and sets the stage for the other three. Changes in the global policy environment bureaucracies find themselves in are understood as a necessary prerequisite for organizational change and adaptation as it is only in response to changing external circumstances and challenges that international bureaucracies come to act in the first place. Each of the four factors has been addressed in some detail by international bureaucracy and public policy scholars and their work lends theoretical weight to the argument. The chapter concludes with the rationale for case selection. The main reason for choosing the UNFCCC Secretariat, IEA Secretariat, and World Bank for the analysis of organizational change and governance integration is their centrality as international bureaucracies within global climate and energy governance architectures. As the leading IFI providing concessional funding for energy infrastructure and climate change mitigation and adaptation projects in the developing economies of the Global South, the World Bank plays an important role in both as well as in the global governance architecture for development.

Chapter 2 begins the core empirical discussion with a focus on the UNFCCC Secretariat and its activities since the early 1990s. As the main forum to diplomatically address global climate change internationally, the climate convention occupies the central role within the global climate governance architecture. The chapter illustrates how the UNFCCC Secretariat transitioned from an earlier, science and environment-focused convention process to one that addresses energy issues much more directly, given the closely intertwined nature of fossil fuel combustion and global climate change. As an international bureaucracy with little autonomy and a comparatively small size initially, the Secretariat has increasingly established itself as a potent "orchestrator," guiding negotiators and more self-assuredly pushing the climate convention process forward. This has not least been due to the entrepreneurial role played by the UNFCCC's Executive Secretaries over time. However, given its statutory limitations, the UNFCCC Secretariat's ability to affect governance integration is limited.

Chapter 3 continues with a closer look at the IEA Secretariat. The world's leading intergovernmental energy organization has a smaller membership than the UNFCCC or the World Bank due to statutory limitations, which also restrict its ability to directly influence policy decisions taken by member states. And yet the world's leading authority on energy statistics is widely accepted to play an influential role in energy and climate change policy debates. The IEA was created to enable its members to develop a coordinated response to oil supply disruptions. Moving well beyond the confines of its original mission, it has, in recent years, undertaken important steps to widen its issue portfolio (in part due to the shock effect of the creation of IRENA), build partnerships with the UNFCCC Secretariat and the IPCC, and advocate for more aggressive climate change mitigation efforts, drawing the global energy and climate governance architectures closer together.

Chapter 4 sheds more light on the activities of the World Bank as the leading IFI supporting the developing countries of the Global South. With its focus on economic development and poverty reduction, the World Bank may not seem, at least initially, to be a key player in energy and climate governance. Yet its essential and growing activities in both fields, including through an increasing alignment with the goals of the climate convention, have changed the role of the Bank fundamentally. The SDGs are a reflection of a new reality in international development that acknowledges climate change and energy as key dimensions that need to be addressed in tandem. The activities of the World Bank are, today, unthinkable without a significant focus on the interconnected issue areas. This chapter explains how these changes came to pass, showing how the World Bank transformed itself from a strong focus on concessional lending for fossil fuel infrastructure in developing countries and comparatively little concern for environmental impacts in its infant days, into a player in full recognition of its indispensable role in driving the twenty-first century's interconnected global climate and sustainable development agendas.

The concluding chapter explores both empirical and normative implications of the argument. The two main findings consistent across the three cases are, first, that international bureaucracies have been undergoing periods of organizational change that, second, have provided the basis for an integration of climate and energy governance architectures. The analytical framework proposes an explanation for this change and the discussion of the changing activities of the UNFCCC Secretariat, the IEA Secretariat, and the World Bank illustrates how this has played out in practice. However, the findings raise questions about the role of international bureaucracies and the future of global climate and energy governance. What are the effects of organizational change in international bureaucracies? What are the effects of governance integration? Does it produce specific outcomes that move the world forward? Should we want to see more integration? With the climate emergency impacting on all policy fields, integrating it into the structure and activities of international bureaucracies not traditionally operating specifically in the climate change space seems apt and, indeed, necessary if effective solutions are to be accessible to and understood by as large an audience as possible. Similarly, incorporating a greater focus on energy and other related issue areas into the UNFCCC is only logical given the close interlinkages between energy production and consumption, and anthropogenic climate change. The final chapter pulls together the key findings of this study and assesses the meaning of governance integration between global climate and energy governance architectures moving forward.

Notes

1 Stephen Browne and Thomas G. Weiss, "A UN Fit for Purpose?" *New Discourse*, (London: UNA-UK, 2016). www.sustainablegoals.org.uk/wp-content/uploads/2016/03/152-154-WEISS_NEW-Discourse_2016_Revise.pdf.
2 UNFCCC, Paris Agreement (Bonn: UNFCCC, 2015). http://unfccc.int/files/essential_background/convention/application/pdf/english_paris_agreement.pdf.

3 For an in-depth discussion of the Industrial Revolution, emissions trajectories, and the onset of anthropogenic climate change see John L. Brooke, *Climate Change and the Course of Global History: A Rough Journey* (Cambridge: Cambridge University Press, 2014).

4 The 1850 data is drawn from the World Resources Institute's (WRI) CAIT 2.0 climate data explorer. For 2019 emissions date, see IEA, *Global Energy & CO2 Status Report 2019* (IEA, Paris, 2019). www.iea.org/reports/global-energy-co2-status-report-2019.

5 IEA, *Energy and Climate Change* (Paris: IEA, 2015). www.iea.org/reports/energy-and-climate-change.

6 IRENA, *Renewable Capacity Statistics 2020* (Abu Dhabi: IRENA, 2020). www.irena.org/-/media/Files/IRENA/Agency/Publication/2020/Mar/IRENA_RE_Capacity_Statistics_2020.pdf.

7 IPCC, *Global Warming of 1.5°C. An IPCC Special Report on the Impacts of Global Warming of 1.5°C above pre-industrial Levels and related global Greenhouse Gas Emission Pathways, in the Context of strengthening the global Response to the Threat of Climate Change, Sustainable Development, and Efforts to eradicate Poverty* (Geneva: IPCC, 2018). www.ipcc.ch/site/assets/uploads/sites/2/2019/06/SR15_Full_Report_High_Res.pdf.

8 UNFCCC, *Kyoto Protocol to the United Nations Framework Convention on Climate Change* (Kyoto: UNFCCC, 1997). https://unfccc.int/sites/default/files/resource/docs/cop3/l07a01.pdf.

9 Due to a lack of interest in climate change and broader environmental concerns and, relatedly, an absence of environment ministries or equivalent, some countries (e.g., OPEC members) were for many years represented only through their energy and resource ministries.

10 Harald Heubaum and Frank Biermann, "Integrating Global Energy and Climate Governance: The Changing Role of the International Energy Agency," *Energy Policy* 87 (2015), 229–239.

11 Kate O'Neill, *The Environment and International Relations* (Cambridge: Cambridge University Press, 2009).

12 See, e.g., Thomas Bernauer, "The Effect of International Environmental Institutions: How We Might Learn More," *International Organization* 49, no. 2 (1995), 351–377; Detlef F. Sprinz and Carsten Helm, "The Effect of Global Environmental Regimes: A Measurement Concept," *International Political Science Review* 20, no. 4 (1999), 359–369; Oran Young, *International Cooperation: Building Regimes for Natural Resources and the Environment* (Ithaca, N.Y.: Cornell University Press, 1989); and Jørgen Wettestad, *Designing Effective Environmental Regimes: The Key Conditions* (Cheltenham, UK: Edward Elgar, 1999).

13 See, e.g., Frank Biermann et al., eds., *International Organizations in Global Environmental Governance* (London: Routledge, 2009); Robert Falkner, *Business Power and Conflict in International Environmental Politics* (Basingstoke, UK: Palgrave Macmillan, 2008); Sebastian Oberthür and Olaf S. Stokke, eds., *Managing Institutional Complexity: Regime Interplay and Global Environmental Change* (Cambridge: MIT Press, 2011); Philipp

Pattberg, *Private Institutions and Global Governance: The New Politics of Environmental Sustainability* (Cheltenham, UK: Edward Elgar, 2007); and Paul Wapner, "Politics Beyond the State: Environmental Activism and World Civic Politics," *World Politics* 47, no. 3 (1995), 311–340.

14 See, e.g., Frank Biermann et al., eds., *Global Climate Governance Beyond 2012. Architecture, Agency and Adaptation* (Cambridge: Cambridge University Press, 2010); Harriet Bulkeley, *Accomplishing Climate Governance* (Cambridge: Cambridge University Press, 2015); Joyeeta Gupta, *The History of Global Climate Governance* (Cambridge: Cambridge University Press, 2014); David Held et al., *Climate Governance in the Developing World* (Cambridge: Polity Press, 2013); and Matthew J. Hoffmann, *Climate Governance at the Crossroads: Experimenting with a Global Response after Kyoto* (Oxford: Oxford University Press, 2011).

15 See, e.g., Frank Biermann et al., "The Fragmentation of Global Governance Architectures: A Framework for Analysis," *Global Environmental Politics* 9, no. 4 (2009), 14–40; Marcel J. Dorsch and Christian Flachsland, "A Polycentric Approach to Global Climate Governance," *Global Environmental Politics* 17, no. 2 (2017), 45–64; Harro van Asselt, *The Fragmentation of Global Climate Governance: Consequences and Management of Regime Interactions* (Cheltenham, UK: Edward Elgar, 2014); Fariborz Zelli and Harro van Asselt, "The Institutional Fragmentation of Global Environmental Governance: Causes, Consequences, and Responses," *Global Environmental Politics* 13, no. 3 (2013), 1–13; and Rüdiger Wurzel et al., eds., *Pioneers, Leaders and Followers in Multilevel and Polycentric Climate Governance* (Abingdon, UK: Routledge, 2020).

16 See, e.g., Morris A. Adelman, "Oil Fallacies" *Foreign Policy* 82 (1991), 3–16; John Surrey, "Japan's uncertain Energy Prospects: the Problem of Import Dependence," *Energy Policy* 2, no. 3 (1974), 204–230.

17 Ann Florini and Benjamin K. Sovacool, "Who Governs Energy? The Challenges Facing Global Energy Governance," *Energy Policy* 37, no. 12 (2009), 5239–5248.

18 See, e.g., Aleh Cherp et al., "Governing Global Energy: Systems, Transitions, Complexity," *Global Policy* 2, no. 1 (2011), 75–88; Andreas Goldthau, "Governing Global Energy: Existing Approaches and Discourses," *Current Opinion in Environmental Sustainability* 3, no. 4 (2011), 213–217; Timothy Meyer, "The Architecture of International Energy Governance," *American Society of International Law Proceedings* 106 (2013), 389–394; Thijs van de Graaf, *The Politics and Institutions of Global Energy Governance* (Houndmills, UK: Palgrave, 2013); and Thijs van de Graaf and Jeff Colgan, "Global Energy Governance: A Review and Research Agenda" *Palgrave Communications* 2 (2016), 1–12.

19 See, e.g., Christian Downie, "Strategies for Survival: The International Energy Agency's Response to a New World," *Energy Policy* 141 (2020), 1–7; Ann Florini, "The International Energy Agency in Global Energy Governance," *Global Policy* 2, no. S1 (2010), 40–50; Thijs van de Graaf, "Obsolete or Resurgent? The International Energy Agency in a Changing

Global Landscape," *Energy Policy* 48 (2012), 233–241; and Thijs van de Graaf and Dries Lesage, "The International Energy Agency after 35 years: Reform Needs and Institutional Adaptability," *The Review of International Organizations* 4, no. 3 (2009), 293–317.

20 See, e.g., Christian Downie, "Global Energy Governance in the G20: States, Coalitions and Crises," *Global Governance* 21, no. 3 (2015), 475–492; and Philip Andrews-Speed and Xunpeng Shi, "What Role Can the G20 Play in Global Energy Governance? Implications for China's Presidency," *Global Policy* 7, no. 2 (2016), 198–206.

21 See, e.g., Stephen P.A. Brown and Hillard G. Huntington, "OPEC and World Oil Security," *Energy Policy* 108 (2017), 512–523; Jeff D. Colgan, "The Emperor Has no Clothes: The Limits of OPEC in the Global Oil Market," *International Organization* 68, no. 3 (2014), 599–632; and Gal Hochman and David Zilberman, "The Political Economy of OPEC," *Energy Economics* 48 (2015), 203–216;

22 See, e.g., Johannes Urpelainen and Thijs van de Graaf, "The International Renewable Energy Agency: A Success Story in Institutional Innovation?," *International Environmental Agreements* 15 (2015), 159–177; Indra Overland and Gunilla Reischl, "A Place in the Sun? IRENA's Position in the Global Energy Governance Landscape," *International Environmental Agreements: Politics, Law and Economics* 18 (2018), 335–350; and Thijs van de Graaf, "Fragmentation in Global Energy Governance: Explaining the Creation of IRENA," *Global Environmental Politics*, 13 (2013), 14–33.

23 See, e.g., Gonzalo Escribano, "Fragmented Energy Governance and the Provision of Global Public Goods," *Global Policy* 6, no. 2 (2015), 97–106; Rafael Leal-Arcas et al., *International Energy Governance: Selected Legal Issues* (Cheltenham, UK: Edward Elgar, 2015); and Van de Graaf, "Fragmentation in Global Energy Governance: Explaining the Creation of IRENA," 14–33.

24 Fiona Harvey, "World Has Six Months to Avert Climate Crisis, says Energy Expert" *The Guardian*, 18 June 2020.

25 Jeff Colgan, "The International Energy Agency: Challenges for the 21st Century" *GPPI Policy Paper Series* 6 (Berlin: Global Public Policy Institute, 2009).

26 For exceptions, see, e.g., Robert Falkner, "Global Environmental Politics and Energy: Mapping the Research Agenda" *Energy Research & Social Science*, 1 (2014), 188–197; Heubaum and Biermann, "Integrating Global Energy and Climate Governance: The Changing Role of the International Energy Agency," 229–239; Benjamin K. Sovacool and Thijs van de Graaf, "Building or Stumbling Blocks? Assessing the Performance of Polycentric Energy and Climate Governance Networks," *Energy Policy* 118 (2018), 317–324; and Fariborz Zelli, "Global Climate Governance and Energy Choices" in *Handbook of Global Energy Policy*, ed. Andreas Goldthau (Chichester, UK: Wiley-Blackwell, 2013), 340–357.

27 See, e.g., Frank Biermann, *Earth System Governance: World Politics in the Anthropocene* (Cambridge: MIT Press, 2014); Frank Biermann

and Rakhyun E. Kim, eds., *Architectures of Earth System Governance* (Cambridge: Cambridge University Press, 2020); and Sarah Burch et al., "New Directions in Earth System Governance Research," *Earth System Governance* 1 (2019), 1–18.

28 Steffen Bauer et al., "Understanding International Bureaucracies: Taking Stock," in *Managers of Global Change: The Influence of International Environmental Bureaucracies*, eds. Frank Biermann and Bernd Siebenhüner (Cambridge: MIT Press, 2009), 15–36.

29 See, e.g., Michael N. Barnett and Martha Finnemore, "The Politics, Power, and Pathologies of International Organizations," *International Organization* 53, no. 4 (1999), 699–732; Michael W. Bauer and Jörn Ege, "Bureaucratic Autonomy of International Organizations' Secretariats," *Journal of European Public Policy* 23, no. 7 (2016), 1019–1037; Robert W. Cox and Harold K. Jacobson, *The Anatomy of Influence. Decision making in International Organizations* (New Haven: Yale University Press, 1973); and Bob Reinalda and Bertjan Verbeek, eds., *Autonomous Policy Making by International Organizations* (London: Routledge, 1998).

30 See, e.g., Steffen Bauer et al., "International Bureaucracies," in ed. Frank Biermann and Phillip Pattberg, *Global Environmental Governance Reconsidered* (Cambridge: MIT Press, 2012), 27–45; and Frank Biermann and Bernd Siebenhüner, eds., *Managers of Global Change: The Influence of International Environmental Bureaucracies* (Cambridge: MIT Press, 2009).

31 See, e.g., Sikina Jinnah, "Marketing Linkages: Secretariat Governance of the Climate-Biodiversity Interface," *Global Environmental Politics* 11, no. 3 (2011), 23–43; and Bob Reinalda and Bertjan Verbeek, "Leadership of International Organizations," in *Oxford Handbook of Political Leadership*, eds., Roderick, A. W. Rhodes and Paul 't Hart (Oxford: Oxford University Press, 2013), 595–609.

32 See, e.g., Frank R. Baumgartner and Bryan D. Jones, *Agendas and Instability in American Politics* (Chicago: University of Chicago Press, 1993); John W. Kingdon, *Agendas, Alternatives, and Public Policies* (Boston: Little Brown, 1984); and Michael Mintrom and Phillipa Norman, "Policy Entrepreneurship and Policy Change" *Policy Studies Journal* 37, no. 4 (2009), 649–667.

33 Ernst B. Haas, *Where Knowledge Is Power: Three Models of Change in International Organizations* (Berkeley: University of California Press, 1990).

34 Paul Sabatier and Hank Jenkins-Smith, *Policy Change and Learning: An Advocacy Coalition Approach* (Boulder, Colo.: Westview Press, 1993).

35 Baumgartner and Jones, *Agendas and Instability in American Politics*

36 Thomas A. Birkland, "Focusing Events, Mobilization, and Agenda Setting," *Journal of Public Policy* 18, no. 1 (1998), 53–74.

37 Alexander L. George and Andrew Bennett, *Case Studies and Theory Development in the Social Sciences* (Cambridge: MIT Press, 2005).

38 It is possible, as Bennett and Checkel argue, that within-case process-tracing may uncover causal mechanisms

that may be either very generalizable or unique to one or a few cases, but it is almost impossible to know prior to researching a case the degree to which any inductively derived explanations will be one or the other.

(Andrew Bennett and Jeffrey T. Checkel, *Process Tracing: From Metaphor to Analytic Tool* (Cambridge: Cambridge University Press, 2014))

39 Peter G. Swanborn, *Case Study Research: What, Why and How?* (London: Sage, 2010).

40 George and Bennett, *Case Studies and Theory Development in the Social Sciences.*

41 Ibid.

42 Bennett and Checkel, *Process Tracing: From Metaphor to Analytic Tool.*

43 Heubaum and Biermann, "Integrating Global Energy and Climate Governance: The Changing Role of the International Energy Agency," 229–239.

1 Studying governance integration
A conceptual framework

This chapter develops the conceptual framework used in this study to analyze processes of organizational change in international bureaucracies as the driving force behind governance integration. It begins by defining governance integration as a discrete and novel concept, following a brief review of the literature on the fragmentation of governance architectures. While governance integration has not received much scholarly attention to date, it is closely related to existing accounts of policy integration and polycentric governance. The chapter then conceptualizes international bureaucracies as the objects of this study and elaborates on what is required for governance integration to occur: organizational change in these bureaucracies prompted by a changing global policy environment. Although organizational change has been addressed by the IR and international bureaucracy literatures, existing definitions of the concept are often labored and in need of greater clarity.

The third main section sets out this book's analytical approach, detailing the four key factors that are argued to explain the change that has taken place: a changing global policy environment, organizational autonomy, organizational structure, and organizational leadership. The choice of factors is adapted from the wider literature on international bureaucracies more generally, and Bauer et al.'s conceptual exploration of organizational change in international bureaucracies and Biermann and Siebenhüner's work on the influence of international environmental bureaucracies more specifically.[1] However, influence is not considered as a separate factor here as the influence of the UNFCCC Secretariat, the World Bank, and the IEA Secretariat, respectively, is treated as a given and as it is not directly relevant to this work to assess how the three international bureaucracies shape policy outcomes in various countries or how the information they disseminate is received by the wider public, the media, and specialized policy communities. This study aims to make

DOI: 10.4324/9781315661339-2

sense of the four factors through the lens of governance integration in an increasingly polycentric governance system. A changing global policy environment, organizational autonomy, organizational structure, and organizational leadership are conceptualized by drawing on international bureaucracy, IR, and public policy literatures. The final section details the rationale for selecting the three cases.

Conceptualization of governance integration

In contrast to governance integration, the fragmentation of governance architectures has received significant scholarly attention in recent years. The consequence has been a detailed theoretical understanding of the concept of fragmentation as well as an extensive empirical application, particularly in the field of global environmental governance. Expanding on Raustiala and Victor's definition of a regime complex as "an array of partially overlapping and non-hierarchical institutions governing a particular issue-area,"[2] Biermann et al. define the term governance architecture as the "overarching system of public and private institutions that are valid or active in a given issue area," that is as comprising "organizations, regimes, and other forms of principles, norms, regulations and decision-making procedures" and resembling "the meta-level of governance."[3] A governance architecture may be considered fragmented, they further contend, when the issue area governed by this architecture is "marked by a patchwork of international institutions that are different in their character (organizations, regimes, and implicit norms), their constituencies (public and private), their spatial scope (from bilateral to global), and their subject matter (from specific policy fields to universal concerns)." A degree of fragmentation has been observed to be common to all global governance architectures, regardless of the issue area.[4]

And indeed, the issue areas under observation in this study are fragmented with a patchwork of international bureaucracies, states, and non-state actors addressing a variety of different aspects. While the global climate governance architecture is more hierarchical than its counterpart in the energy field, with the UNFCCC as the central framework around which most other actors revolve, it is still marked by the presence of a multitude of different organizations at various levels of political authority. These organizations may well partially overlap in their goals and activities, which is hardly surprising given the cross-cutting, global nature of climate and energy concerns. However, for the purposes of this research, the work on fragmentation is only the starting point. Indeed, as Biermann et al. point out, "the concept of architecture allows for the analysis of situations of both synergy and conflict between different regimes or other types of institutions."[5] This study seeks to show how growing synergies may lead to integration within and between governance architectures. While there can be no governance integration without some kind of fragmentation, integration is more than just the interaction of actors in an otherwise fragmented governance environment.

Therefore, despite the close links to governance fragmentation, governance integration is not simply the opposite of the former. It is a discrete and novel concept with an independent quality. Building on Heubaum and Biermann, governance integration is defined as a positive interaction of actors either within a governance architecture or, the focus of this study, between two (or more) different governance architectures that may result in a convergence of approaches and activities.[6] Governance integration, whether within a governance architecture or between two different architectures, may begin to occur if at least one of the actors involved changes or adjusts their approaches and activities in ways that align it with the approaches and activities of another actor or group of actors. In the context of this study, it would require an international bureaucracy operating in global energy or climate governance or, indeed, in a related field, to adapt and start behaving in ways that align its approaches and activities with those of another international bureaucracy, operating either within global energy or global climate governance.

Governance integration is related to the concept of polycentricity as applied to climate change governance by Ostrom[7] and further developed by Jordan et al.,[8] among others. Understood as a polycentric system, global climate governance is a complex network of "multiple governing authorities at different scales rather than a mono-centric unit. Each unit within a polycentric system exercises considerable independence to make norms and rules within a specific domain" in order to effectively address climate change.[9] Such a system exists both vertically, across multiple levels of political authority, from the international all the way down to the local, and horizontally, for example, across the multitude of actors populating this governance arena at the international level, including international bureaucracies as the objects of this study. The core assumptions are not dissimilar to the work on governance fragmentation discussed above; yet in its normative application "on how better to govern, poly-centric governance thinking provides a rather different starting point to other stock-in-trade terms and concepts."[10] Polycentric governance moves beyond the focus on states as governing climate change within the UNFCCC regime complex and avoids the somewhat more negative connotations of governance fragmentation through an acknowledgment of the opportunities for learning, innovation, integration, and mutual reinforcement in global climate change governance that arise from the activities of a range of independent yet connected actors.

The concept is also closely related to the literatures on legal and policy integration and mainstreaming, which have drawn on examples ranging from EU policymaking to international development cooperation.[11] There are a number of important overlaps between legal and policy integration on the one hand and governance integration on the other. First, both assess the ways in which new responsibilities are integrated into separate policy streams, be it in local planning bodies, national government ministries, or policy units within international bureaucracies. For example, manifestations can be found in the integration of EU nature conservation objectives into urban land use planning in Bulgaria,[12] the integration of environmental concerns into national energy and agriculture ministries in Sweden,[13] or, indeed, the integration of climate change matters into the activities of the IEA as a leading international bureaucracy operating in the energy field.[14]

Second, both emphasize a number of requirements necessary for integration to occur as well as criteria against which such integration can be measured. In policy integration studies, this includes the creation of coordinating structures and legal procedures as well as the need for communication and dialogue to explain and relate the interconnections and need for integrated approaches to address external changes.[15] Relatedly, environmental and climate mainstreaming assesses the directed inclusion of environmental and climate considerations into all relevant activities pursued by a state or non-state actor.[16] While there is no one agreed upon definition of mainstreaming, it is here understood as the active promotion, usually by governments, of climate goals and sustainability as cross-cutting issues in the identification, planning, design, and implementation of strategies, policies, and investment programs. As regards governance integration, this study identifies organizational change and adaptation driven by external challenges as the main determinant. This change is evident in a widening of the issue portfolio the bureaucracy addresses which is, in turn, connected to changing internal structures, the building of partnerships and closer interaction with other actors operating within energy and/or climate governance, and a change advocacy on behalf of specific choices which points to the importance of communicating integration efforts.

However, policy integration, mainstreaming, and governance integration are also conceptually different. Policy integration and mainstreaming follow the kind of designed strategy or top-down plan not currently evident across cases of governance integration. For example, within the EU and its member states, integrating formerly separate policy portfolios in one ministry (such as climate change and energy) and, indeed, mainstreaming specific goals and targets into local planning policies are the intended outcomes of deliberate acts by a government. Governance integration as understood in this study, on the other hand, emerges through international bureaucracies' various efforts to pursue and broaden their mandate in a complex and fast-changing global policy environment largely independent of individual governments. Even though member states exert a degree of control, the relative autonomy of the international bureaucracies analyzed herein has meant that their changing activities have not always come as the result of explicit government direction or, indeed, top-down direction within the bureaucracy. Further, policy integration and mainstreaming in supranational, national, and local contexts have direct impact on policymaking taking place at each of these levels of political authority. While international bureaucracies operating within climate and energy governance show some elements of policy integration and mainstreaming, no such direct impact can be established. The UNFCCC Secretariat, the IEA Secretariat, and the World Bank are each influential players in their own right, but they are not directly involved in domestic policymaking processes. However, recent research has demonstrated that international bureaucracies have considerable impact on global governance even without these direct powers.[17] They do so, for example, through the provision of energy policy expertise in the case of the IEA Secretariat, advocacy and agenda setting in the case of the UNFCCC Secretariat, and implementation and financial support in the case of the World Bank. Governance integration as a concept is therefore specific to the context in which it is found.

Conceptualization of international bureaucracies

The empirical examples discussed in this study are the administrative entities of IOs in the case of the IEA Secretariat and the World Bank, and an international treaty in the case of the UNFCCC Secretariat. International bureaucracies, often also referred to as international public administrations, are here understood, following Biermann et al., as

agencies that have been set up by governments or other public actors with some degree of permanence or coherence and beyond formal direct control of single national governments (notwithstanding control by multilateral mechanisms through the collective of governments) and that act in the international arena to pursue a policy.[18]

This definition separates international bureaucracies from a range of other international actors, including transnational non-governmental organizations and transnational corporations. It also distinguishes them from IOs.

IOs are generally understood to also possess the physical presence that marks international bureaucracies, but conceptually they are more encompassing than the latter. Various IR accounts have defined IOs as specific institutional arrangements with a normative framework, a state membership, and a permanent secretariat to carry out administrative tasks.[19] States set up IOs and delegate some decision-making authority to the administrative entity with the expectation of reducing the transaction costs of cooperation, though the degree to which authority is delegated can vary greatly.[20] Intergovernmental organizations as particular kinds of IOs are further "established by treaty and usually, in order to safeguard state sovereignty, operate at the level of consent, recommendation, and cooperation rather than through compulsion or enforcement."[21] But while IOs are, critically, defined through their state members and the benefits these organizations provide in meeting members' interests, international bureaucracy research has focused more specifically on the actions of "international secretariats" which make up the administrative arms of IOs.[22] International bureaucracies are directly connected to but also separate from, and to some degree autonomous of, the group of member states that created the IOs they serve in the first place.[23] The fact that no single national government can exert sole control over their activities makes international bureaucracies both public and non-state actors.[24] This conceptual distinction allows for international bureaucracies to be independently studied and analyzed as increasingly self-determining actors within global governance architectures. The organizational change in those international bureaucracies is the main focus of this study.

Conceptualization of organizational change

Governance integration is herein argued to occur as the consequence of organizational change within international bureaucracies that is, in turn, shaped by a fast-changing global policy environment. Organizational innovation and change are subjects of a vast literature, especially in management studies and sociology.[25] Hage defines organizational innovation as "the adoption of an idea or behaviour that is new to the organization."[26] This innovation is further thought to consist of the three elements of "change, novelty and improvements in performance" within the organization. Change itself is a constant and organizations need to continually renew their "direction, structure, and capabilities to serve the ever-changing needs of external and internal customers."[27] Whilst a useful start, the management literature's focus on innovation and changes in business practices in private, for-profit organizations is limited in its applicability to IOs and their bureaucracies such as the ones discussed in this study. This is because of clear differences in the target structure (quantitative versus qualitative targets), accountability (shareholders versus states and other IOs), autonomy (independent decisions-making versus limited freedom to act), and high versus low adaptability to change.[28]

As discussed earlier, the IR literature has historically understood changes in IOs as primarily the result of changing state preferences. In contrast, more recent accounts have emphasized internal organizational dynamics as driving factors for organizational change, including in international bureaucracies. However, extant research has done surprisingly little to explain what it means by organizational change. Haas's work distinguishes between incremental organizational change defined as "the successive augmentation of an organization's program as actors add new tasks to older ones without any change in the organization's decision-making dynamics or mode of choosing"; and turbulent change where said changes in decision-making are present.[29] In a similar fashion, Rochester discusses organizational tinkering and rethinking as forms of organizational change in the UN, describing the former as "mere nuts-and-bolts changes (which would not energize serious involvement in institution-building and would amount to rearranging the deckchairs on the Titanic)" and the latter as "a sweeping rewriting of the Charter (which seems too titanic an endeavor)."[30] In assessing the outcome of activities of international bureaucracies and the behaviors of actors within them, Biermann et al. employ the concept of "relative change" which is based on a comparison of actor behavior before and after the organizational activity. Acknowledging potential problems with their approach, the authors' aim is to compare outcomes against a counterfactual situation in order "to assess an improvement in relation to the hypothetical state of affairs that would most likely not have occurred in the absence of activity of the bureaucracy in question."[31]

Each of these accounts offers somewhat cumbersome definitions of organizational change. Most of the literature, however, does not even go so far but rather assumes organizational change to be an accepted and understood concept without offering a succinct definition. It is useful, then, to compare the international bureaucracy literature's approach to that taken in other related disciplines in which change is a core concept. The public policy literature widely and succinctly understands policy change to be "an alteration in the commitment of a government to an objective," that is, a change in the way governments address and manage policy problems, usually on the basis of regulatory and legislative change.[32] Renewable energy policy may serve as an example. A government may see the amount of GHGs emitted domestically as a problem, what with its international climate change commitments and potential economic benefits of transitioning toward a low-carbon economy. Given that most GHG emissions are generated through energy production and consumption, the energy sector and the policies governing it would be the most obvious choice for change. If a government's objective is to expand electricity generation from low-carbon renewable sources such as wind or solar photovoltaic, the alteration in commitment may come in the form of new legislation (e.g., feed-in tariffs, generation targets, other support schemes, a reduction in fossil fuel subsidies) to provide the new market entrants with an opportunity to challenge entrenched energy sources such as coal and gas and their associated generation infrastructures. Such changes break with established patterns of governance, often producing new ways to deal with perceived problems. In that same vein, individuals in an international bureaucracy focused on energy policy may, for whatever their reasons, come to see climate change as a key issue the bureaucracy should address. With this as an objective, the alteration in commitment may come through the setting up of a new unit within the international bureaucracy, more funding assigned to work on climate change-related topics, or greater advocacy on behalf of climate action internally as well as externally.

Drawing on the concise definition provided by public policy studies and adjusting it to suit the purpose of analyzing change in international bureaucracies in the energy and climate space, organizational change is thus defined herein as the change in the commitment of an international bureaucracy to an objective, resulting in the adoption of new approaches and activities. It is these approaches and activities that this study sets out to trace and analyze for the UNFCCC Secretariat, the IEA Secretariat, and the World Bank. The following section will add more detail to each one of them.

Organizational change

The analysis shows that there are a number of observable sets of activities that are common to the cases studied herein and that may well extend to other international bureaucracies beyond the scope of this research. These activities include the broadening of the issues and projects each international bureaucracy pursues (both internally and externally) and the ways in which it pursues them, the building of partnerships and closer cooperation with other stakeholders within energy and/or climate governance, and a sustained change advocacy on behalf of specific actions or policy choices. Taken together, these activities represent the kind of organizational change required for governance integration to occur between the global energy and climate governance architectures which, while closely related, have remained as if in separate silos for much of the last three decades.

First, a broadening of the issues and projects an international bureaucracy pursues and changes in the ways in which it pursues them are often tied to internal structural changes. For example, new units addressing climate change or renewable energy sources may be created and more staff may be recruited to bring the new structure to life. New financial instruments and positions may be developed to support specific goals. Regardless of the structural changes, however, the international bureaucracy, through its expert staff, may engage in discussions on issues heretofore unaddressed or add projects and criteria to its work that shift the focus of the bureaucracy. A widening issue portfolio may also be reflected in the form of reports focusing specifically on new issues or incorporating them to a greater extent into an existing publication series. Additional issues may be addressed following the integration of objectives such as the goals of the climate convention into a bureaucracy's operations.

Second, partnerships and closer interaction with other stakeholders in the international system are both an important consequence of as well as parallel development to a broadening of the issues and projects an international bureaucracy pursues. Increased collaboration may be with a variety of different actors, though this study focuses mainly on the interaction between international bureaucracies and organizations as opposed to states. This is in large part because the international bureaucracies analyzed herein already have (with the exception of the IEA) a globally encompassing state membership. Further, closer partnerships and cooperation may be formed with other international bureaucracies and organizations in the same issue area or between two related issue areas, following the definition of governance integration as a positive interaction of actors either within a governance architecture or between two (or more) different governance architectures that may result in a convergence of approaches and activities. Governance integration could also occur without the formation of closer partnerships between different international bureaucracies, for example, if bureaucracies simply changed their activities along similar lines to those pursued elsewhere without more direct interaction, in effect achieving a more aligned and integrated approach to governance. However, given the complex interconnections between energy and climate change on the one hand and actors in the international system on the other, this is an unlikely development.

Finally, the strong advocacy on behalf of policy choices aimed at mitigating and adapting to climate change and transitioning to a low-carbon future is the third key organizational change observed and it is the one with arguably the most far-reaching consequences. First, like the building of closer partnerships and cooperation with other actors in the field, successful external change advocacy is dependent on and runs parallel to changes internal to the bureaucracy. But while the widening of the issue portfolio, often through the integration of key objectives, and the building of closer partnerships with other international actors may already expand the remit of the international bureaucracy, advocating a strong change message is, second, akin to a qualitative change that further advances a sense of organizational autonomy. Through their entrepreneurial emphasis on the importance of change and the choices that should flow from the realization that the climate emergency needs to be urgently addressed, the leadership team of the three international bureaucracies, each of which is the preeminent international authority in its field, are having an impact on the global energy and climate policy debate and the variety of actors within it.

Change factors

In their discussion of UN peace operations, Bauer et al. attempt to understand organizational change in international bureaucracies by pointing to a number of determinants, including the organizational environment and the influence of member states as external factors, and the nature of the organization, size, leadership, internal politics, organizational culture, identity, and reform attempts as internal factors.[33] Analyzing the changing role and influence of international environmental bureaucracies, Biermann and Siebenhüner employ an analytical framework that also incorporates a number of external and internal factors, including organizational autonomy, embeddedness, organizational structure, organizational culture, people and procedures, resources, organizational leadership, and the nature of the problem the bureaucracies exist to address.[34] In their application of Biermann and Siebenhüner's approach to analyze the autonomous influence of the UN Division of Sustainable Development, Widerberg and van Laerhoven argue against overly complex and detailed approaches and for improving the framework by usefully collapsing the unnecessarily large number of factors while emphasizing their interplay.[35] Heeding this call, this study builds on both Bauer et al. and Biermann and Siebenhüner in its development of a set of focused internal and external change factors that in their interplay are argued to account for the organizational change observed. These factors, or variables, may be categorized as the global policy environment, organizational autonomy, organizational structure, and organizational leadership.

Each can be linked to a rich scholarly tradition in the study of international bureaucracies, IR, and public policy studies. It is important to note here that the first variable, the global policy environment, critically sets the stage for the other three. Changes in the global policy environment in which bureaucracies operate are a key prerequisite for organizational change and adaptation as it is only due to changing circumstances and greater external challenges creating the need to act that international bureaucracies come to change in the first place. In many ways, the global policy environment, although itself subject to constant change, is here seen as a static variable when compared to the more dynamic and interconnected variables of organizational autonomy, structure, and leadership. Aspects such as growing scientific evidence in support of anthropogenic climate change or the dramatic decline in the cost of renewables, to name but two, are the same across the cases analyzed in this study. However, there is some variation in the way that the UNFCCC Secretariat, the IEA Secretariat, and the World Bank respond to a changing global policy environment and which of its aspects they emphasize both internally and in their engagement with other stakeholders.

The global policy environment

First, the cases analyzed in this book embed international bureaucracies in a complex and fast-changing global policy environment that requires them to adapt if they want to remain relevant in energy and climate change debates. Within the extant literature, there are a number of useful accounts to help delimit the global policy environment as a concept. Public policy scholars have pointed to the importance of the social, political, and economic context in influencing policymaking processes.[36] For example, Sabatier and Weible differentiate between stable and dynamic factors external to the subsystems in which policymaking occurs. Stable factors "include basic attributes of the problem (e.g., the difference between groundwater and surface water), the basic distribution of natural resources, fundamental sociocultural values and structure, and basic constitutional structure," while "dynamic external factors include changes in socioeconomic conditions, changes in the governing coalition, and policy decisions from other subsystems."[37] In contrast to these predominantly domestic factors, the policy networks approach aims to incorporate transnational and supranational dimensions, also referred to as "internationalized policy environments."[38] Much of the attention on such environments has focused on the role of the EU and its influence in setting the stage for decision-making in member states.[39] Transnational contexts are seen to "serve as a macropolitical opportunity structure that adds new opportunities and constraints for domestic actors."[40]

These opportunities may be found in what Kingdon has termed "policy windows."[41] Policy windows are "opportunities for action on given initiatives, [which] present themselves and stay open for only short periods of time." They disrupt the policy equilibrium and offer policy entrepreneurs the chance to push and advocate for policy change. Policy windows may be triggered by the outcome of a national election, the emergence of pressing policy problems (such as climate change), or by a crisis or focusing event such as a major power plant accident. Focusing events in particular are seen to offer a distinct opportunity for policy innovation and change. To Birkland, a focusing event is a "sudden, relatively rare" occurrence that is harmful or suggests future harm in a "definable geographical area or community of interest, and that is known to policymakers and the public virtually simultaneously."[42] Analyzing watershed management by local conservation authorities in the Canadian Province of Ontario, Michaels et al. show how three focusing events – a hurricane, drastic cuts in government funding for conservation efforts, and water supply contamination – created policy windows for practitioners to take action in increasing source water protection.[43] However, while focusing events may "reinforce some preexisting perception of a problem," they often do not in themselves carry that problem to a prominent position on a policy agenda.[44]

It is clear, then, that exogenous factors, be they slower or more dramatic developments, have impact on policymaking processes. Within public policy studies, this holds for research into agenda setting as well as into policy networks. Changes to the policy environment

> can affect the resources, interests and relationships of the actors within networks. Changes in these factors can produce tensions and conflicts which lead to either a breakdown in the network or the development of new policies. However, these changes do not have

an effect independent of the structure of, and interactions within, the network.[45]

In other words, changing social, political, and economic conditions are necessary but insufficient. They do not drive change in and of themselves. For this, other factors are also required.

Similar to public policy studies but locating their focus on international dimensions, IR scholars have attempted to position policy issues within a global context. Reinicke situates policymaking as becoming "increasingly influenced by global conditions," specifically the "growing social and economic integration around the world [which] has extended the geographic scope of public policy far beyond national borders."[46] As the localized release of CO_2 and other GHGs has brought about the carbon-loading of the atmosphere and anthropogenic climate change with its truly globalized implications, so addressing them requires global policy solutions, too. In her discussion of global public policy and transnational policy communities, Stone employs the idea of a "global agora" as a "growing global public space of fluid, dynamic and intermeshed relations of politics, markets, culture and society" within which global policy processes may give rise to just such solutions.[47] These processes are themselves shaped by a variety of domestic, international, and transnational actors. Stone's work builds on other previous scholarly efforts, including Dryzek's idea of a "global public sphere"[48] and Ronit and Schneider's discussion of a "global arena"[49] within which international actors operate. However, the global policy environment remains difficult to conceptualize as most research in IR and other related disciplines takes the existence of such an environment for granted without defining it more fully.

This study understands the global policy environment as an environment encapsulating policy issues and related factors that transcend national boundaries and necessitate responses by a global community which, in turn, includes a wide range of actors such as states, NGOs, businesses, the media, as well as IOs and their bureaucracies. This policy environment is different from the legal, institutional, and financial framework that states as the bureaucracies' principals have set for them and which Biermann et al. refer to as "polity."[50]

The global policy environment in the climate and energy fields is shaped by a number of aspects, including scientific evidence without which global climate change would not be recognized as a problem in the first place, diplomatic efforts to find collective solutions to mitigate and adapt to climate change, a power shift in the global system from North to South that requires broader, more inclusive efforts at climate control, a host of climate and energy policies enacted in a number of different jurisdictions around the world, the changing economics of renewable energy and other low-carbon alternatives to fossil fuels, and the growing awareness of climate change risks across the public and private sectors which further broadened the field of stakeholders. Taken together, these more macro aspects both push and pull international bureaucracies toward incorporating a greater number of issues into their portfolios and addressing climate change and energy more holistically.

For example, scientific evidence on the reality of anthropogenic climate change has been a part of global policy discourses in the environmental field for decades. Although the global phenomenon was first discovered as early as the late nineteenth century by Swedish scientist Svante Arrhenius, it did not rise to greater prominence on both domestic and international policy agendas until the 1970s and 1980s when long-term measurements of the concentration of CO_2 in the Earth's atmosphere provided the more reliable data needed to demonstrate the causal links to human energy use. In 1977, the US National Academy of Sciences issued a stark report warning that due to coal combustion, global average surface temperatures could rise by several degrees above pre-industrial levels by the middle of the twenty-first century. Internationally, scientists began drawing together their findings at the first World Climate Conference organized by the World Meteorological Organization (WMO) in Geneva in 1979.

The creation of the IPCC nine years later, following a period of agenda setting at the domestic level, including through a series of US Congressional hearings on climate change organized by then-Congressman Al Gore in the early 1980s, can be seen as a watershed moment in the relationship between science and the policymaking and diplomatic processes. The IPCC, originally created by the WMO and the United Nations Environment Programme (UNEP), has since produced six major reports assessing the state of scientific evidence on the physical basis of anthropogenic climate change, its mitigation as well as its impacts and adaptation. These reports, published in 1990, 1995, 2001, 2007, 2014, and 2021/2022 respectively, assessing and citing countless scientific papers and reports and drawing on the contributions of thousands of leading experts, have supported and in many ways enabled the work of the UNFCCC Secretariat, other non-state actors and states pushing for urgent action to address anthropogenic climate change as a global emergency. The increasingly large amount of physical evidence on the reality of anthropogenic climate change has also made it much more difficult to deny or ignore the problem. The IPCCs AR5 states that warming of the atmosphere and ocean system is "unequivocal" and that human activity is "extremely likely" to have been the dominant cause of this additional warming, further increasing the level of certainty from the preceding report.[51] Like other IPCC reports before, it focused the attention of governments, the media, international bureaucracies, businesses, environmental NGOs, and other stakeholders, setting the stage for negotiations at the Twenty-First Conference of the Parties (COP21) to the UNFCCC in Paris, for more action at national and subnational levels, and for greater integration between the climate and related governance architectures.

Organizational autonomy, structure, and leadership

The dimensions of the global policy environment represent the main axes along which change in the global climate and energy space has occurred. They have developed incrementally as scientific evidence, diplomatic progress, power shifts, domestic policies, economic cost decreases, as well as climate risk awareness have accumulated over time. Together, they provide both the definition of the problem (dangerous anthropogenic climate change) and its solution (a low-carbon transition based on an international agreement that, in turn, depends on supportive domestic policies, affordable technology solutions, and public and private sector financial support). As such, they have a consequential impact on the actions and behaviors of states and subnational governments, businesses, and investors, as well as IOs and their bureaucracies. In other words, the global policy environment drives changes internal to a variety of different stakeholders operating in the energy and climate policy space.

Organizational change in international bureaucracies – that is, the change in the commitment of these bureaucracies to an objective – resulting in the adoption of new approaches and activities, can be affected by changes in the global policy environment in a variety of ways. For one, the international bureaucracy, through its leadership and staff at different levels of authority, may pick up on changes in any of the four dimensions and translate them into changing understandings, behaviors, and activities. For an international bureaucracy like the UNFCCC Secretariat, the state of climate science, the translation of international goals and ambition into domestic legislative and executive practice, and heightened awareness of climate change as an issue in the private sector are all critical in lending support to the Secretariat's efforts. They provide a beneficial setting for the expansion of the international bureaucracy, changes in structure, cooperation with other stakeholders, and, relatedly, external change advocacy. But changing behaviors and activities may also be the result of attempts to adapt to more direct challenges to the international bureaucracy itself. For the IEA Secretariat, for example, the creation of IRENA in 2009 as, at least initially, a rival organization acted as a catalyst to widen its issue portfolio and pay greater attention to renewables and the transition to a low-carbon energy system of the future.[52] The creation of IRENA came as a consequence of the growing and more competitive role of renewables which some IEA member states didn't consider to be effectively represented or, indeed, fairly assessed by an Agency too focused on fossil fuels.[53] Since 2009, the IEA has intensified its work on renewables and low-carbon transition pathways, both independently of and in cooperation with IRENA.

These and other developments draw greater attention to the role of states in shaping global governance architectures and they raise questions over the relationship between international bureaucracies and the IOs' member states on whose behalf the bureaucracy is conducting its business. The following section will shed more light on issues of organizational autonomy and the interplay between national governments as principals and international bureaucracies as their agents.

Organizational autonomy

Research on international bureaucracies has found them to be influential actors in a variety of different policy fields. They perform a number of functions, from creating and maintaining regimes to the delivery of global services such as disarmament verification or trade dispute resolution.[54] If international bureaucracies are to effectively act in and adapt to a changing global policy environment, they need to be able to flexibly adjust their approaches and activities in accordance with these changes. However, the extent to which they are able to do so depends on the autonomy afforded to them by the IOs' member states. Much of the research on international bureaucracies since the 1970s has focused on this autonomy and what bureaucracies make of it.[55] Seen through the prism of principal-agent theory, states (the principals) created IOs and their bureaucracies (the agents) to perform certain functions benefiting them as a collective, for example, through coordination or information gathering and dissemination amongst all members of the IO.[56]

Much of the earlier application of principal agent approaches to the study of IOs and their bureaucracies focused on the role of the state principal and the ways in which principals would be able to control agents that were understood to have a "certain minimal autonomy

[as] a natural product of delegation."[57] More recent work has stressed the ability of international bureaucracies to become more autonomous from their principals, emphasizing the need to understand such agents as actors in their own right, and finding them to be dynamic rather than "static instruments of intergovernmental policymaking."[58] This has been found to be especially the case in situations of information asymmetry and conflicting interests between principals and agents that may create "agency slippage between what the principal wants and what the agent does."[59] This slippage may be enhanced and the agent may become more autonomous if there is a collective principal made up of multiple actors as delegation amongst these actors is more complicated and their policy preferences are not always homogenous.[60]

All three international bureaucracies observed in this study have a group of states as their collective principal. In the case of the UNFCCC, member states are represented as parties to the convention, an international treaty process with ongoing, annual negotiations. IEA member states exercise control via the Governing Board. They influence the direction of the World Bank through the Board of Governors and the Board of Executive Directors. The UNFCCC Secretariat has the largest number of states making up its principal, closely followed by the World Bank and then, with some distance, the IEA Secretariat.[61] Policy preferences are indeed multifarious in each case, with regard to the UNFCCC and the World Bank more so than the IEA given its smaller membership made up of developed OECD economies only. However, despite having a larger membership than the other IOs, the UNFCCC Secretariat, serving as the international bureaucracy to the climate convention, is not the most autonomous of the cases studied herein. This is because of limitations thrown up by its narrower and more specific mandate. The World Bank's broad mandate on the other hand has given it much greater independence and decision-making authority vis-à-vis its collective principal. And yet, each of the three international bureaucracies has exercised its mandate in novel ways and attempted to broaden it over time.

Given the observed changes in each of the three cases discussed herein, it is clear that a degree of autonomy and independence enables changes in activities to be driven in unexpected ways: from within the bureaucracy rather than simply through influence or pressure from states. The UNFCCC Secretariat, the IEA Secretariat, and the World Bank all operate with (varying) degrees of autonomy from the states that set them up and/or constitute their membership, making them useful case studies in gaining an understanding of organizational change and governance integration. And yet autonomy alone is an insufficient explanation for change. International bureaucrats may well find themselves in an environment affording them greater freedom to make independent decisions but this tells us little about whether or not they choose to do so, which actors in the bureaucracy pursue which course of action, to which challenges they might react, and precisely how they would go about introducing new and innovative approaches to the activities of the bureaucracy.

Organizational structure

International bureaucracies perform a number of functions for the IOs they serve, including gathering information, organizing meetings, and developing a variety of different projects and programs. These functions are carried out by both permanent and temporary staff, including officials seconded from member states. Importantly, however, rather than serving the interests of individual state principals, the primary loyalty of staff rests with the international bureaucracy.[62] Research has found staff to develop and pursue a range of different interests, legitimizing their work through a fulfillment of the organization's mandate.[63] Further, international bureaucracies have been found to exhibit a "compound nature consisting of multiple behavioural dynamics, role definitions and identities."[64] Differentiated specializations and responsibilities within the bureaucracy, thus, allow for the pursuit of a number of different agendas under the same organizational roof.

In their study of bureaucratic autonomy of international organization secretariats, Bauer and Ege incorporate structural features of these secretariats into a more encompassing conceptualization of organizational autonomy. In addition to "the ability of international secretariats to develop autonomous bureaucratic preferences (autonomy of will)," structural changes within international secretariats are expressed in "their capacity to transform these preferences into action (autonomy of action)."[65] The international bureaucracies observed in this study do have the ability to decide on staff hires, the issues staff focus on within the bureaucracy, as well as the setup of topical units, although significant growth in capacity also depends on a number of external factors, not least the budget provided by member states over time. This study treats organizational structure as closely related to but separate from organizational autonomy. Doing so allows for a more detailed assessment of internal structural changes, such as the creation of new units within the bureaucracy or new senior management roles addressing climate change, and their role in driving organizational change and, consequently, governance integration.

Organizational leadership

As Burns noted over four decades ago, "leadership is one of the most observed and least understood phenomena on earth."[66] Since then, research in the social sciences has attempted to make sense of leaders in public, private, and non-governmental organizations and the impact their actions have on changes within these organizations and beyond. The literature on international bureaucracies has identified organizational leadership as a key factor in explaining why and how international bureaucracies change and wield influence in the international arena. Special emphasis has been given to leaders' and, consequently, the international bureaucracies' ability "to adapt their goals, internal processes, and the organizational structure to perceived external challenges."[67] External challenges arising as the result of a complex and dynamic global policy environment discussed above need to be translated both within the bureaucracy and in the bureaucracies' outward engagement with external actors. However, beyond the finding that strong organizational leadership positively correlates with organizational performance and influence, international bureaucracy scholarship has not devoted sustained attention to analyzing the entrepreneurialism displayed by senior actors within the bureaucracy. In each of the cases discussed in this study, senior bureaucrats have been found to be critically important to the processes of organizational change. The public policy literature's treatment of policy entrepreneurs driving policy change offers useful insights to help make more sense of this dynamic.

Policy entrepreneurs play a vital role in driving policy change by "attempting to alter other people's understandings of the issues with which they deal."[68] They make deliberate efforts to shape policy images to capture attention and gain support for their solutions. The route to success often runs through the media, where "the set of images of public issues put forward [...] is determined by a mix of factual circumstances and by the interpretations attached to these circumstances by policy entrepreneurs."[69] Bauer calls this entrepreneurial strategy "discourse framing," which is defined as a "subtle [...] strategy to influence the interpretation of the problem, thereby pre-determining possible answers."[70] In other words, policy entrepreneurs are critically important in determining problem definitions.

But policy entrepreneurs do not only work on defining or re-defining problems and ideas. They also advocate for the use or disuse of particular policy instruments based on their interests, perceptions, and ideology. Entrepreneurs can be interest groups, research organizations, elected or appointed government officials, or, as in this study, senior civil servants in an international bureaucracy. What matters is that they have a way to wield influence as well as a willingness to do so. As Mintrom and Norman point out, most of the actors participating in policy-making processes "are comfortable working within established institutional arrangements; doing their bit to achieve improved outcomes for themselves and their supporters without upsetting the status quo." What sets policy entrepreneurs apart is "their desire to significantly change current ways of doing things in their area of interest."[71] They identify and advocate on behalf of specific policy solutions, revealing themselves "through their attempts to transform policy ideas into policy innovations and, hence, disrupt status quo policy arrangements."[72] Policy entrepreneurs often don't act alone but seek to maximize their impact by cooperating with others, in this study either within or outside the international bureaucracy, to successfully challenge dominant paradigms.[73] This is crucial considering the urgent need for the world to shift away from emissions-intensive, business-as-usual pathways toward sustainable development to effectively address the climate emergency and a host of other global challenges facing humanity in the twenty-first century.

As the three cases analyzed in this study show, senior leaders, including the Executive Secretaries and Deputy Executive Secretaries of the UNFCCC Secretariat, the Executive Directors and Chief Economists of the IEA Secretariat, the Presidents and Vice Presidents and heads of particular units of the World Bank, acted entrepreneurially in helping to drive internal changes and externally position the international bureaucracy through defining problems and solutions, setting agendas, using windows of opportunity, building coalitions beyond their governance fields, and, as a consequence, drawing the climate and energy governance architectures closer together.

Case selection

The scholarly work produced on international bureaucracies to date has analyzed a wide range of cases. However, due to the large number of IOs and secretariats in existence today as well as the associated difficulty of producing in-depth examinations for each of them, the majority of publications has focused on small-N analyses, selecting only the most visible and influential cases. For example, Barnett and Finnemore draw on the International Monetary Fund (IMF), the Office of the United Nations High Commissioner for Refugees (UNHCR), and UN Peacekeeping to illustrate the changing role, authority, and autonomy of international secretariats.[74] Biermann and Siebenhüner assemble research on nine leading international bureaucracies active in the environmental field, citing the need for "intense qualitative analysis" as a reason for restricting the number of case studies chosen.[75]

This study in organizational change and governance integration follows previous approaches for three reasons. First, it is interested in the changing activities of international bureaucracies over time and the factors that have led to this change. Producing empirical evidence for the key factors identified requires an in-depth tracing of developments through interviews and documents which naturally limits the number of case studies that can effectively be included. Second, the goal of the cases studied is not to derive directly transferable results, although the analytical approach that underpins the research can be applied to other cases. The factor determining the number of cases chosen is, therefore, not their number but the depth and quality of within-case analysis. Third, the focus is on integration between global energy and climate governance architectures as two specific fields. Each of the three international bureaucracies analyzed occupies a key role within one or both of those architectures. The UNFCCC Secretariat hosts the international negotiation process and climate convention around which global climate governance revolves; the IEA is the leading authority within global energy governance; and the World Bank is the leading international financial institution providing financial support to both energy and climate change-related projects in developing countries of the Global South and is a key player in both architectures. This makes the three cases interesting and highly relevant.

All three cases represent different types of international bureaucracies. The UNFCCC Secretariat is the product of an international treaty process. Since the early 1990s, its main role has been to support and shepherd international climate change negotiations but in recent years its activities have extended along with a growth in size, an inflow of greater and more varied technical expertise, and a more active leadership by its Executive Secretaries. The Secretariat is linked to the UN but is different from its established programs. Its changing importance within the UN family is reflected in the role of the Executive Secretary which was recently elevated to the level of Under Secretary General with the Deputy Executive Secretary now holding the rank of Assistant Secretary General. As a case, the UNFCCC Secretariat is interesting from a number of perspectives, especially regarding organizational leadership, organizational autonomy, changes in structure, and the bureaucracies' efforts to respond to a changing global policy environment.

As a component of the World Bank Group, the World Bank is part of the UN system. However, the world's leading international financial institution operating in the Global South is also different from the more established arms of the UN, including its various programs. The creation of the World Bank at the 1944 Bretton Woods Conference preceded the creation of the UN by more than a year and the size of the Bank's lending for projects in developing countries makes it the largest multilateral development bank (MDB) operating in the Global South. Environmental protection was incorporated into the policy objectives of the World Bank in the 1980s and there has been a dramatic shift in the attitude of the bureaucratic arm of the Bank toward climate change issues from a more unwelcoming approach to a full embrace of the issue in recent years. During this time, the World Bank's work on energy issues developed further as well, with a growing focus on providing concessional finance for low-carbon, climate-friendly projects. With its shifting attitudes to climate change as an issue and the establishment and growth of units and financial instruments within the bureaucracy over time, the case of the World Bank lends further weight to the factors identified in the analytical framework.

Finally, the IEA was established in 1974 as an autonomous organization within the framework of the Organisation for Economic Cooperation and Development (OECD). In addition to its original role of helping member states coordinate measures during oil supply disruptions, the IEA Secretariat has provided policy advice on a growing range of energy and environmental issues, has helped train officials in national ministries of both member states and partner countries, and has, through its leadership, increasingly advocated energy policy pathways in line with global climate change mitigation targets. The shift from an international bureaucracy focused exclusively on oil concerns toward one addressing all aspects of the energy system, incorporating climate change into its portfolio, cooperating closely with the UNFCCC, and advocating for a rapid decarbonization of the global energy system to tackle the climate emergency, makes the IEA another significant case in which to analyze organizational change and the integration between global energy and climate governance architectures.

Notes

1 Frank Biermann and Bernd Siebenhüner, eds., *Managers of Global Change: The Influence of International Environmental Bureaucracies* (Cambridge: MIT Press, 2009).
2 Kal Raustiala and David G. Victor, "The Regime Complex for Plant Genetic Resources," *International Organization* 58, no. 2 (2004), 277–309.

3 Frank Biermann et al., "The Fragmentation of Global Governance Architectures: A Framework for Analysis," *Global Environmental Politics* 9, no. 4 (2009), 14–40.

4 Fariborz Zelli and Harro van Asselt, "The Institutional Fragmentation of Global Environmental Governance: Causes, Consequences, and Responses," *Global Environmental Politics* 13, no. 3 (2013), 1–13.

5 Biermann et al., "The Fragmentation of Global Governance Architectures: A Framework for Analysis," 14–40.

6 Harald Heubaum and Frank Biermann, "Integrating Global Energy and Climate Governance: The Changing Role of the International Energy Agency," *Energy Policy* 87 (2015), 229–239; See also Serge M. Garcia et al. (eds.), *Governance of Marine Fisheries and Biodiversity Conservation: Interaction and Coevolution* (Chichester: John Wiley & Sons, 2014).

7 Elinor Ostrom, "Polycentric Systems for Coping with Collective Action and Global Environmental Change," *Global Environmental Change* 20, no. 4 (2010), 550–557.

8 Andrew Jordan et al., eds., *Governing Climate Change. Polycentricity in Action?* (Cambridge: Cambridge University Press, 2018).

9 Ostrom, "Polycentric Systems for Coping with Collective Action and Global Environmental Change," 550–557.

10 Jordan et al., eds., *Governing Climate Change. Polycentricity in Action?*

11 See, e.g., William Lafferty and Eivind Hovden, "Environmental Policy Integration: Towards an Analytical Framework," Environmental Politics 12, no. 3 (2003), 1–22; Andrea Lenschow, ed., *Environmental policy integration: Greening sectoral policies in Europe* (London: Earthscan, 2002); and Joyeeta Gupta and Nicolien van der Grijp, eds., *Mainstreaming Climate Change in Development Cooperation: Theory, Practice and Implications for the European Union* (Cambridge: Cambridge University Press, 2010).

12 Vanja Simeonova and Arnold van der Walk, "The Need for a Communicative Approach to Improve Environmental Policy Integration in Urban Land Use Planning," *Journal of Planning Literature* 23, no. 3 (2009), 241–261.

13 Mans Nilsson and Katarina Eckerberg, eds., *Environmental Policy Integration in Practice: Shaping Institutions for Learning* (London: Earthscan, 2007).

14 Heubaum and Biermann, "Integrating Global Energy and Climate Governance: The Changing Role of the International Energy Agency," 229–239.

15 Vanja Simeonova and Arnold van der Walk, "Environmental Policy Integration: Towards a Communicative Approach in Integrating Nature Conservation and Urban Planning in Bulgaria," *Land Use Policy* 57 (2016), 80–93.

16 See, for example, Hens Runhaar et al., "Mainstreaming Climate Adaptation: Taking Stock about "What Works" from Empirical Research Worldwide," *Regional Environmental Change* 18 (2018), 1201–1210.

17 Dominique De Wit et al., "International Bureaucracies," in *Architectures of Earth System Governance: Institutional Complexity and Structural*

Transformation, eds. Frank Biermann and Rakhyun E. Kim (Cambridge: Cambridge University Press, 2020), 57–74.

18 Frank Biermann et al., "Studying the Influence of International Bureaucracies: A Conceptual Framework," in *Managers of Global Change: The Influence of International Environmental Bureaucracies,* eds. Frank Biermann and Bernd Siebenhüner (Cambridge: MIT Press, 2009), 37–74.

19 Ibid. See also Bob Reinalda, ed., *Routledge Handbook of International Organization* (Abingdon: Routledge, 2013).

20 Joel P. Trachtman, "The Economic Structure of the Law of International Organizations," *Chicago Journal of International Law* 15, no. 1 (2014), 162–194.

21 Alvin L. Bennett and James K. Oliver, *International Organizations: Principles and Issues,* 7th edition (Upper Saddle River, NJ: Prentice Hall, 2001).

22 John Mathiason, *Invisible Governance: International Secretariats in Global Politics* (Bloomfield, CT: Kumarian Press, 2007). See also Bob Reinalda and Bertjan Verbeek, eds., *Decision Making within International Organizations* (London: Routledge, 2004).

23 Michael W. Bauer and Jörn Ege, "Bureaucratic Autonomy of International Organizations' Secretariats," *Journal of European Public Policy* 23, no. 7 (2016), 1019–1037; see also Daniel L. Nielson and Michael J. Tierney, "Delegation to International Organizations: Agency Theory and World Bank Environmental Reform," *International Organization* 57, no. 2 (2003), 241–276.

24 Steffen Bauer and Silke Weinlich, "International Bureaucracies: Organizing World Politics," in *The Ashgate Research Companion to Non-State Actors*, ed. Bob Reinalda (Farnham: Ashgate, 2011), 251–262.

25 See, e.g., Fariborz Damanpour, "Organizational Innovation: A Meta-Analysis of Effects of Determinants and Moderators," *The Academy of Management Journal* 34, no. 3 (1991), 555–590; Fariborz Damanpour and Marguerite Schneider, "Phases of the Adoption of Innovation in Organizations: Effects of Environment, Organization and Top Managers," *British Journal of Management* 17, no. 3 (2006), 215–236; Barbara Levitt and James G. March, "Organizational Learning," *Annual Review of Sociology* 14 (1988), 319–340; and Marshall S. Poole and Andrew H. van de Ven, *Handbook of Organizational Change and Innovation* (Oxford: Oxford University Press, 2004).

26 Jerald T. Hage, "Organizational Innovation and Organizational Change," *Annual Review of Sociology* 25 (1999), 597–622.

27 John W. Moran and Baird K. Brightman, "Leading Organizational Change," *Career Development International* 6, no. 2 (2001), 111–118.

28 Steffen Bauer et al., "Understanding International Bureaucracies: Taking Stock," in *Managers of Global Change: The Influence of International Environmental Bureaucracies*, eds. Frank Biermann and Bernd Siebenhüner (Cambridge: MIT Press, 2009), 15–36.

29 Ernst B. Haas, *Where Knowledge Is Power: Three Models of Change in International Organizations* (Berkeley: University of California Press, 1990).

30 J. Martin Rochester, *Waiting for the Millennium: United Nations and the Future of World Order* (Columbia: University of South Carolina Press, 1993).
31 Biermann et al., "Studying the Influence of International Bureaucracies: A Conceptual Framework," 37–74.
32 Bryan D. Jones and Frank R, Baumgartner, *The Politics of Attention: How Government Prioritizes Problems* (Chicago: University of Chicago Press, 2005); see also Frank R. Baumgartner and Bryan D. Jones, *Agendas and Instability in American Politics* (Chicago: University of Chicago Press, 1993); Michael Howlett et al., *Studying Public Policy: Policy Cycles and Policy Subsystems*, 3rd edition (Oxford: Oxford University Press, 2009); Peter John, "Is There Life After Policy Streams, Advocacy Coalitions, and Punctuations: Using Evolutionary Theory to Explain Policy Change?," *Policy Studies Journal* 31, no. 4 (2003), 481–498.
33 Steffen Bauer et al., "Organizational Change in International Bureaucracies," in *The Management of UN peacekeeping: Coordination, Learning, and Leadership in Peace Operations*, eds. Julian Junk et al. (Boulder, CO.: Lynne Rienner, 2017), 239–264.
34 Biermann and Siebenhüner, eds., *Managers of Global Change: The Influence of International Environmental Bureaucracies.*
35 Oscar Widerberg and Frank van Laerhoven, "Measuring the Autonomous Influence of an International Bureaucracy: The Division for Sustainable Development," *International Environmental Agreements* 14 (2014), 303–327.
36 Howlett et al., *Studying Public Policy: Policy Cycles and Policy Subsystems.*
37 Sabatier, Paul, and Christopher M. Weible, "The Advocacy Coalition Framework: Innovations and Clarifications," in *Theories of the Policy Process,* ed. Paul A. Sabatier (Boulder, CO: Westview Press, 2007), 189–222.
38 William D. Coleman and Anthony Perl, "Internationalized Policy Environments and Policy Network Analysis," *Political Studies* 47, no. 4 (1999), 691–709.
39 See, e.g., John Peterson, "Decision-Making in the European Union: Towards a Framework for Analysis," *Journal of European Public Policy* 2, no. 1 (1995), 69–93.
40 Silke Adam and Hanspeter Kriesi, "The Network Approach," in *Theories of the Policy Process,* ed. Paul A. Sabatier (New York: Routledge, 2007), 129–154.
41 John W. Kingdon, *Agendas, Alternatives, and Public Policies* (Boston: Little Brown, 1984).
42 Thomas Birkland, *After Disaster: Agenda-Setting, Public Policy, and Focusing Events* (Washington, DC: Georgetown University Press, 1997).
43 Sarah Michaels et al., "Policy Windows, Policy Change, and Organizational Learning: Watersheds in the Evolution of Watershed Management," *Environmental Management* 38 (2006), 983–992.
44 Kingdon, *Agendas, Alternatives, and Public Policies.*
45 David Marsh and Martin Smith, "Understanding Policy Networks: Towards a Dialectical Approach," *Political Studies* 48, no. 1 (2000), 4–21.

46 Wolfgang H. Reinicke, "The Other World Wide Web: Global Public Policy Networks," *Foreign Policy* 117 (1999), 44–57.
47 Diane Stone, "Global Public Policy, Transnational Policy Communities, and Their Networks," *Policy Studies Journal* 36, no. 1 (2008), 19–38.
48 John S. Dryzek, "Transnational Democracy," *The Journal of Political Philosophy* 7, no. 1 (1999), 30–51.
49 Karsten Ronit and Volker Schneider, eds., *Private Organisations in Global Politics* (London: Routledge, 2000).
50 Biermann and Siebenhüner, eds., *Managers of Global Change: The Influence of International Environmental Bureaucracies.*
51 IPCC, *Climate Change 2014: Synthesis Report: Contribution of Working Groups I, II and III to the Fifth Assessment Report of the Intergovernmental Panel on Climate Change* (Geneva: IPCC, 2014). IPCC reports use {this insertion changes the meaning of the sentence – please remove, I have deleted here} a likelihood scale to assign a level of confidence for each finding. An extremely likely outcome has a probability of 95–100 percent.
52 Heubaum and Biermann, "Integrating Global Energy and Climate Governance: The Changing Role of the International Energy Agency," 229–239.
53 Thijs Van de Graaf, "Fragmentation in Global Energy Governance: Explaining the Creation of IRENA," *Global Environmental Politics* 13, no. 3 (2013), 14–33.
54 Mathiason, *Invisible Governance. International Secretariats in Global Politics.*
55 See, e.g., Michael N. Barnett and Martha Finnemore, "The Politics, Power, and Pathologies of International Organizations," *International Organization* 53, no. 4 (1999), 699–732; Bauer and Ege, "Bureaucratic Autonomy of International Organizations' Secretariats," 1019–1037; Robert W. Cox and Harold K. Jacobson, *The Anatomy of Influence. Decision Making in International Organization* (New Haven: Yale University Press, 1973); and Bob Reinalda and Bertjan Verbeek, eds., *Autonomous Policy Making by International Organizations* (London: Routledge, 1998).
56 See, e.g., David Lake et al., "The Logic of Delegation to International Organizations," in *Delegation and Agency in International Organizations,* eds. Darren G. Hawkins et al. (Cambridge: Cambridge University Press, 2006).
57 Manfred Elsig, "Principal–Agent Theory and the World Trade Organization: Complex Agency and 'Missing Delegation,'" *European Journal of International Relations* 17, no. 3 (2010), 495–517.
58 Bauer et al., "Organizational Change in International Bureaucracies," 239–264.
59 Nielson and Tierney, "Delegation to International Organizations: Agency Theory and World Bank Environmental Reform," 241–276; see also Jonas Tallberg, "The Anatomy of Autonomy: An Institutional Account of Variation in Supranational Influence," *Journal of Common Market Studies* 38, no. 5 (2000), 843–864.

60 Ibid.; see also Darren G. Hawkins et al., eds., *Delegation and Agency in International Organizations* (Cambridge: Cambridge University Press, 2006).

61 A total of 197 parties ratified the UNFCCC, including all UN member states plus non-members Niue and the Cook Islands, non-member observer state Palestine, and the EU. The Paris Agreement on Climate Change has also been ratified by 197 country parties. The World Bank is split into two main parts: the International Bank for Reconstruction and Development (IBRD) with 189 members and the International Development Association with 173 member states. Compared to this, the IEA is small with only 30 members. However, its OECD-only membership includes all developed economies. It also has eight associated members and three countries currently seeking membership: Chile, Israel, and Lithuania.

62 Jarle Trondal, "International Bureaucracy: Organizational Structure and Behavioural Implications," in *Routledge Handbook of International Organization,* ed. Bob Reinalda (Abingdon, UK: Routledge, 2013), 162–175.

63 Susan H. Allen and Amy T. Yuen, "The Politics of Peacekeeping: UN Security Council Oversight across Peacekeeping Missions," *International Studies Quarterly* 58, no. 3 (2014), 621–632.

64 Jarle Trondal et al., *Unpacking International Organisations: The Dynamics of Compound Bureaucracies* (Manchester, UK: Manchester University Press).

65 Bauer and Ege, "Bureaucratic Autonomy of International Organizations' Secretariats," 1019–1037.

66 James M. Burns, *Leadership* (New York: Harper & Row, 1978).

67 Biermann et al., "Studying the Influence of International Bureaucracies: A Conceptual Framework," 37–74.

68 Baumgartner and Jones, *Agendas and Instability in American Politics.*

69 Ibid.

70 Michael W. Bauer, "Limitations to Agency Control in European Union policy-making: The Commission and the Poverty Programmes," *Journal of Common Market Studies* 40, no. 3 (2002), 381–400.

71 Michael Mintrom and Phillipa Norman, "Policy Entrepreneurship and Policy Change," *Policy Studies Journal* 37, no. 4 (2009), 649–667.

72 Michael Mintrom, "So You Want to Be a Policy Entrepreneur?" *Policy Design and Practice* 2, no. 4 (2019), 307–323.

73 Sander Meijerink and Dave Huitema, "Policy Entrepreneurs and Change Strategies: Lessons from Sixteen Case Studies of Water Transitions around the Globe," *Ecology and Society* 15, no. 2 (2010), 21–39.

74 Michael Barnett and Martha Finnemore, *Rules for the World: International Organizations in Global Politics* (Ithaca, NY: Cornell University Press, 2004).

75 Biermann and Siebenhüner, eds., *Managers of Global Change: The Influence of International Environmental Bureaucracies.*

2 The transformation of the UNFCCC Secretariat

Geneva, the summer of 1995: Canadian government officials are paying a visit to the newly established UNFCCC Secretariat under its Executive Secretary Michael Zammit Cutajar, trying to establish what the Secretariat is planning in the run-up to COP1 at the end of the year in Berlin. They find a "charming and relaxed" atmosphere but are disappointed by what they perceive to be a lack of understanding of the high politics surrounding the climate change issue. The sense of "going to someone's cottage" does not exactly fill these officials with confidence that the Secretariat will be able to effectively support the forthcoming negotiations and deal with the many diplomatic challenges yet to come.[1]

Paris, 12 December 2015: French Foreign Minister Laurent Fabius brings down his gavel to conclude COP21. The result is the most ambitious and most encompassing climate deal ever struck. Around the conference hall, delegates representing 195 negotiating parties rise for a standing ovation. They are applauding Fabius for his chairing and each other for achieving a real breakthrough, but they are also paying their respects to the UNFCCC and its Executive Secretary Christiana Figueres who, through the forging of new alliances and a strong change advocacy, set the stage for an historic agreement to arise from the disappointment of earlier COPs – an agreement that, if followed through, would markedly change the world's energy landscape.

The two decades leading up to the Paris Agreement were a period of immense change for the UNFCCC Secretariat. From humble beginnings in Geneva, the international bureaucracy transformed itself into a potent force, guiding negotiators and pushing the climate convention process ever forward from its Bonn headquarters. Its staff and budget grew bigger, its Executive Secretaries bolder, and its links with players outside the core climate governance architecture stronger. Against a loose chronological backdrop of the various outcomes – and

DOI: 10.4324/9781315661339-3

often setbacks – of the climate convention process which began in earnest with COP1 in Berlin in December 1995, this chapter traces these changes through an emphasis on several interrelated developments in an international bureaucracy at the heart of, arguably, the most complex and thorny global issue of our time. These developments include changes in the issue portfolio addressed by the Secretariat and the number of expert staff working at headquarters, cooperation with a growing number of international organizations and corporate players outside of the global climate governance architecture, and changing approaches by UNFCCC Executive Secretaries to shape negotiations and frame the wider climate policy debate taking place at international level and in capitals around the world. Taken together, these changes within the UNFCCC Secretariat, including their external effects, have contributed to and paved the way for the process of governance integration analyzed in this book. They provide the basis for the integration of climate change goals into the IEA and the World Bank as the leading bureaucracies in their respective governance architectures. The chapter begins with a look at the role of science in creating the rationale for negotiations to limit GHG emissions and the formation of the UNFCCC Secretariat.

Growing pains

The climate convention process is rooted in – and critically dependent on – the availability of adequate scientific evidence, the gathering of which preceded international negotiations by several decades, beginning in the 1950s with the continuous measurement of atmospheric CO_2 concentrations at the Mauna Loa Observatory, Hawaii. Following moves in the 1970s and 1980s to incorporate scientific findings into the domestic policymaking process, especially in the United States, the creation of the IPCC in 1988 marked a step change in the role of climate science at an international level, institutionalizing an evidence-based process under the auspices of the UN involving thousands of scientists and subject matter experts as well as government representatives of more than 120 countries. Originally set up by the WMO and UNEP at the request of UN member states, the IPCC's main task has since been to prepare and regularly publish scientific assessments of global climate change and its impacts around the world. These so-called Assessment Reports (ARs), published every five to seven years, quickly established themselves as the definitive word on the state of climate science, with every newly released report containing both ever more evidence for the reality of anthropogenic climate change and greater certainty about its causes.[2]

The First Assessment Report (AR1), published in 1990, served as the basis of the UNFCCC that was adopted at the United Nations Conference on Environment and Development (UNCED) in Rio de Janeiro (the "Rio Earth Summit") two years later and entered into force in 1994, following ratification by the Fiftieth Party to the Convention. The UNFCCC Secretariat began its work in 1994 based in Geneva as an autonomous body under the UN Secretary General. Its first Executive Secretary, Michael Zammit Cutajar, had been heading the ad hoc secretariat supporting the intergovernmental negotiations that established the UNFCCC. Now he found himself at the helm of the permanent Convention Secretariat with the rank of UN Assistant Secretary General. Cutajar, an experienced international civil servant, had occupied a number of roles in the UN system, most prominently at the UN Conference on Trade and Development. He had further built credibility in environmental matters as part of Maurice Strong's team preparing and running the 1972 UN Conference on the Human Environment in Stockholm. A year later he followed Strong to the UNEP for a brief stint at the newly created program after the former's elevation to its Executive Secretary.

And yet years later, at the start of his work at the UNFCCC Secretariat, Cutajar "was only vaguely aware of the climate change issue," having joined the process not to make a difference in the fight against human-induced global warming but for the opportunity to build a good team of people and leave behind a well-functioning international bureaucracy.[3] This seemed a tall order given the improvised and humble beginnings illustrated by a first staff meeting fitting "more or less around a dinner table."[4] However, the move to the Secretariat's new headquarters in Bonn, following an offer by the German government to host the UNFCCC, provided a greater opportunity to do so. Geneva was the second largest of the four major office sites of the UN, with permanent missions based there alongside the IPCC and the WMO. As the former capital of West Germany, Bonn offered excellent infrastructure and was no stranger to the international scene. Yet remaining close to UN headquarters was a more interesting proposition for a number of the Secretariat's staff who subsequently stayed behind. The move to Bonn enabled Cutajar to fill gaps and "professionalize the Secretariat with more experts," selecting people not just considering geography (Western powers, China, and Russia had to be represented for political reasons) and, latterly, gender, but also knowledge and fit, at times turning down unsuitable personnel proposals from member countries.[5] From a climate science perspective, hiring new expert staff turned out to be a necessity as the dialogue between the IPCC and the UNFCCC had not been clearly defined at the outset and the UNFCCC

Secretariat often proved unsuccessful in working out precisely what the observations of the IPCC meant with regard to the climate convention process and how this process could resonate more publicly in a way akin to the IPCC reports.[6]

In those early days, similar to other comparable bodies in the UN family, the Secretariat did not play a particularly prominent role. The Executive Secretary and his team "didn't have much of a profile" and did not actively pursue an agenda aimed at shaping policy choices among parties to the convention.[7] This was not just due to the Secretariat's small size and its status as a new kid on the block still trying to establish itself, but also due to an initial view among member states that its purpose was to facilitate negotiations and smooth the process rather than become too visible and independent a player. While there was a need for an international bureaucracy to coordinate activities, it was to be a process driven by states, based on best available science provided by the IPCC, and not by the UN itself.[8] The Secretariat's function was to be seen and not heard, especially in the US's eyes, with additional pushback coming from China and India when an early attempt was made to publish through the UNFCCC, a document compiling a number of proposals for a more effective convention.[9] In interactions with member governments or, indeed, more senior UN bureaucrats, Secretariat representatives would "usually only speak when spoken to." While there was behind the scenes support from various stakeholders for a more active approach to reducing carbon emissions, Secretariat representatives "would get into trouble if they [publicly] talked about carbon or specific industrial sectors, especially energy."[10]

From the UNFCCC's infant days, the apparent conflict between addressing energy as the main driver of anthropogenic climate change, on the one hand, and states' unwillingness to have their domestic energy policy interfered with in a collective attempt to address the problem, on the other hand, impacted negatively on the climate convention.[11] A reading of its first major outcome, the Kyoto Protocol, reveals energy mentioned only six times (of which twice in the Annex); there is no single mention of fossil fuels or coal, and oil and gas are only mentioned once in Annex A.[12] More importantly than individual mentions, however, energy is not discussed in a context establishing direct causality between fossil fuel combustion and the carbon-loading of the atmosphere, a process Kyoto sought to slow if not bring to a halt altogether. Energy was addressed "in a very implicit way primarily because the convention and later the Kyoto Protocol leave it entirely to the states to decide on how mitigations in the various sectors – including energy – are achieved."[13] In addition, given the absence of an overall temperature target in the

Protocol, there is, further, no sense of an upper limit of carbon dioxide that could potentially be emitted before the target cannot be met any longer, something later defined as the global carbon budget. Without such a global budget, and the multitude of national budgets it breaks down into, the energy transition pathways and concrete energy policy choices needed to achieve it are much more difficult to visualize and develop, with all the negative consequences this has for effective climate and energy governance integration.

Governments' concerns over any direct mentions of energy, largely due to a combination of fears over economic impacts, domestic political opposition, and intense lobbying efforts by industry and energy companies such as Exxon Mobil, which concurrently pursued an extensive public misinformation campaign aimed at sowing doubt in the public and policymakers' minds about the reality and scientific underpinnings of anthropogenic climate change as presented by the IPCC and other scientific bodies,[14] provide a convincing and oft-presented yet also incomplete account of this lack of integration. First, the global policy environment had not yet developed in ways advantageous to connecting the dots more directly as part of the climate convention process. Governments needed more time to digest what was being said in scientific reports, greater knowledge and expertise on climate change had yet to be built up in capitals from Washington DC to Beijing, and early impacts of climate change had yet to become visible around the world to help inject a greater sense of urgency into negotiations. Climate legislation and renewable energy policy experiments had yet to succeed in a number of jurisdictions and wider policy learning had yet to occur, economically viable alternative energy choices had yet to fully emerge, and businesses had yet to get behind efforts to regulate carbon emissions. Finally, the links between climate change and energy had yet to be more explicitly established in communicative efforts undertaken by stakeholders across the climate and energy governance architectures. Unlike the Paris Agreement nearly two decades later, negotiations to the Kyoto Protocol did not benefit from having these puzzle pieces in place.

Second, and relatedly, negotiations under the UNFCCC were initially led by representatives from environment and foreign ministries, due to the perception of the issue as primarily environmental in nature on the one hand and the intergovernmental dimension of the treaty that was to be agreed on the other. A broadening of expertise and ministerial representation, to include more finance and energy experts for example, did not occur until much later. Third, despite the critical links between energy and anthropogenic climate change, the IPCC, upon whose reports the convention process and with it the UNFCCC Secretariat are built, was not effective in talking about energy in a way that the energy industry could understand.[15] This would change only gradually over the course of several consecutive ARs and it would take the UNFCCC Secretariat some time to begin forming closer links with established IOs and their bureaucracies in the energy space such as the IEA in order to try and remedy this problem. But before it could even begin this process, it had to deal with a major setback.

Shifting sands

After a difficult start for the UNFCCC and its Secretariat, it is perhaps surprising that the climate convention process produced a tangible result, in the form of the Kyoto Protocol, within just two years after the start of serious negotiations at COP1 in Berlin. To some, it was an "accidental outcome rather than by design."[16] Yet science also had a large role to play. The virtual consensus on the reality of anthropogenic climate change presented in the IPCCs AR2 published in 1995 "was convincing enough for Europe and Japan" to throw their weight behind negotiations in Kyoto two years later.[17] In the end, all major advanced economies supported the Kyoto Protocol. All except the United States, the then world's largest GHG emitter, in a move that threatened to spell doom for the treaty given the United States' global political and economic dominance.

The United States was the first country to ratify the UNFCCC by unanimous Senate consent in the summer of 1992 but this initially supportive role changed at COP1, when developing countries insisted that an agreement to the convention include GHG emissions targets based on historic emissions, laying the responsibility for global climate change and action to deal with its consequences on the doorstep of advanced economies only. The vague, voluntary nature of the UNFCCC had been appealing. Yet the full application of the principle of common but differentiated responsibilities (CBDR) at the COP3 in Japan two years later meant, in effect, that the Kyoto Protocol would set binding targets for industrialized countries to be reached within the compliance period 2008–2012, but not for developing countries like China and India. Furthermore, the Protocol specified abatement targets that were seen as unachievable and cost inefficient.[18] The effects of Kyoto would fall hardest on the energy sector and energy-intensive industries like steel which had been in decline since the late 1970s and 1980s, losing ever more ground to low-wage producers in emerging and developing economies. It came as no surprise then that those standing to incur costs in a low-carbon energy transition rallied in opposition to the Kyoto

Protocol. The Global Climate Coalition (GCC) was formed in 1989, following the creation of the IPCC, as an international interest group of major fossil fuel users and producers with the purpose of lobbying against any moves to mitigate climate change. Kyoto threatened to put actual meat on the UNFCCC's bones and triggered a flurry of activity from the GCC which included ExxonMobil, British Petroleum, the American Petroleum Institute, the National Coal Association, and the world's largest multinational car makers. The GCC fought ongoing international collective action efforts not only by reminding government officials in advanced economies of the potentially economically damaging consequences of reducing carbon emissions but also by questioning the science at the heart of the IPCC and, by extension, the climate convention process. In the US Senate, the GCC found both "especially sympathetic ears and voices as well as those afraid of the political consequences of supporting Kyoto."[19]

In July 1997, before the final negotiations on the Kyoto Protocol had even begun, Democratic Senator Robert Byrd and Republican Senator Chuck Hagel co-sponsored a resolution, which passed the chamber by a vote of 95-0, declaring that the US Senate would oppose a treaty that did not require "new specific scheduled commitments to limit or reduce greenhouse gas emissions for Developing Country Parties within the same compliance period."[20] The Senate resolution mirrored the views of the GCC in its rejection of "serious harm" to the US economy. Although he signed the treaty, US President Bill Clinton never sent it on to the Senate for ratification. In 2001, the George W. Bush administration declared the Kyoto Protocol "dead,"[21] arguing in line with the Byrd-Hagel Resolution that abiding by its goals and accepting CBDR would put the US economy at a competitive disadvantage. At home, the administration refused to consider CO_2 a pollutant to be regulated under the Clean Air Act and blocked California's request for a waiver to impose stricter automobile emissions standards. It did, however, pursue a new national energy policy overseen by Vice President Richard Cheney. As reported by the *New York Times* in 2007, during the drafting of the policy, representatives of dozens of energy-industry groups, largely representative of the later disbanded GCC, met with the Vice President and his staff, successfully injecting their views into the policy and asserting them over environmental concerns in the process.[22]

The UNFCCC's first Executive Secretary could only look on as the politics of climate change threatened to undermine the viability of the treaty. And yet underneath the very public, often antagonistic contestation over how best to proceed in light of an American refusal to participate in the global effort, his goal of building the Secretariat into a functioning international bureaucracy had made quiet progress. When Cutajar left the UNFCCC in 2002, the Kyoto Protocol was still three years from coming into effect; yet staff at the Secretariat had grown, along with a growing budget, from the initial somewhat randomly assembled handful to 200, adding significantly greater expertise and analytical capacity.[23] His successor, Joke Waller-Hunter, seamlessly continued this process, having gained experience as the UN's first Director for Sustainable Development, followed by a stint as Head of the OECD's Environment Directorate. Her comparatively brief and noiseless period in charge of the UNFCCC Secretariat was marked by efforts to consolidate the Secretariat's position and achieve the entry into force of the Kyoto Protocol, a goal achieved in early 2005 following Russia's ratification a few months prior.

With Kyoto came a renewed sense that climate change had to be urgently mitigated – through changes in the energy system – and that adaptation wasn't going to be enough in the face of dramatic consequences, some of which had already started to materialize. Both the IPCC and the UNFCCC Secretariat began to advance a more focused issue definition along these lines, helping to more clearly establish the energy side of the equation as part of the climate convention process.[24] They were aided in this effort by the UK government which at the time was headed by a particularly proactive Prime Minister who had early on understood the challenge of mitigating climate change and was intent on putting the issue front and center.[25] Following the entry into force of the Kyoto Protocol, Tony Blair did so by making climate change one of the two core themes of the 2005 G8 Summit hosted by the United Kingdom in Gleneagles, Scotland. Concluding the summit, leaders issued a joint communiqué and a plan of action explicitly linking the use of fossil energy and anthropogenic climate change, calling for a clean energy transition, and reaffirming their commitment to the UNFCCC.[26] While not binding in nature, the Gleneagles Plan of Action spelled out specifics with regard to energy efficiency, support for renewables, mobilizing clean energy finance, and managing climate change impacts. Importantly, the summit helped "focus the climate debate on energy and the low-carbon transition" which was beginning to bear fruit in Germany and a few other countries with cost reductions for wind and solar PV technology.[27]

In hindsight, the year 2005 marked the opening of a window of opportunity for climate change as the issue rose on the international agenda following the entry into force of the Kyoto Protocol and the G8 Summit. And a new opportunity was coming to the UNFCCC Secretariat, too. Following Waller-Hunter's sudden passing in late 2005, UN Secretary General Kofi Annan chose Yvo de Boer as the

Secretariat's new head. Hailing from the same Dutch ministerial background as his predecessor, De Boer had developed an inside view of climate negotiations through his involvement in shaping the European Union's position in the run-up to the Kyoto Protocol as well as leading delegations to the climate convention process. Unlike Waller-Hunter, de Boer sought to build a greater case for private sector involvement in UNFCCC negotiations, partnering with the World Business Council on Sustainable Development and launching an international dialogue on the Clean Development Mechanism (CDM) involving financial players. His approach put him in contrast to a Secretary General who wasn't particularly interested in climate change and did not proactively advance major initiatives on the issue.[28] But de Boer had no intention of keeping quiet, "making coffee and photocopies" and overseeing day-to-day efforts from the relative isolation of the UNFCCC's headquarter in Bonn.[29] He considered himself less a bureaucrat and more an advocate, intent on raising the profile of climate change as an issue and bringing a new, more proactive style to the office of Executive Secretary and with it the Secretariat. Under de Boer, the Secretariat gradually moved from what has been described as a "technocratic bureaucracy" to something more akin to an "activist bureaucracy" intent on promoting its own agenda.[30]

The publication of the *Stern Review on the Economics of Climate Change* tied in with de Boer's plan of broadening stakeholder participation and it gave him early ammunition in the struggle to build support for attempts to forge an effective path forward to follow on from Kyoto. The Blair government had continued its efforts begun at Gleneagles with the release in 2006 of the *Stern Review* which quickly established itself as the most comprehensive economic assessment of climate change impacts and the action necessary to mitigate them available anywhere. Specifically, it showed that decisive action taken to reduce GHG emissions in the present would far outweigh the costs of doing so at a later date in the future, making the economic case for an ambitious global climate treaty and associated domestic policies.[31] The report reframed the debate as one that was as much about economics as about science and lit a fuse under a climate convention process at the time still largely populated by environmental activists, a situation de Boer himself viewed as problematic. Instead of leaving the field only to "climate warriors and worriers" who, perhaps inadvertently, contributed to the compartmentalization of climate change issues, the goal was to broaden constituencies and instill a sense in negotiations as well as the wider policy debate that climate change and economic growth belonged together and need not be at odds.[32] The core problem preventing international collective action had to be addressed to pave the way for a new agreement and a key to unlocking the issue was to forge closer links with business and financial players, an important step toward integrating related governance architectures.

In the run-up to COP13 in Bali, de Boer met with Indonesia's Minister of Finance Sri Mulyani Indrawati to persuade her to organize a meeting of ministers of finance and economics to be held in parallel to the climate negotiations. The resulting Bali breakfasts were co-chaired by the President of the World Bank Robert Zoellick and the Chair of the International Monetary Fund (IMF) Dominique Strauss-Kahn. The meetings focused on climate change – one of the first concerted high-level efforts to specifically link the climate convention process with financial and economic considerations, albeit initially more from the perspective of government and international financial institutions (IFIs) than that of private and institutional investors. De Boer now had generous buy-in from UN Secretary General Ban Ki-moon who took over from Annan a few months after de Boer was installed at the UNFCCC. As the entre-preneurial de Boer worked to push the issue further up the agenda, Ban helped to ensure national capitals paid sufficient attention.[33]

The year 2007 was to prove a pinnacle of international activity on climate change, with a number of focusing events setting the scene: the IPCC and Al Gore were jointly awarded the Nobel Peace Prize for their work to build knowledge on climate change, on the initiative of the UK government the UN Security Council debated climate change as a security threat for the first time in its history representing a further broadening of the debate, and the roadmap developed and agreed at COP13 in Bali spelled out a clear path for a new global agreement to be completed two years later at COP15 in Copenhagen. Yet far from the crowning achievement of de Boer's tenure at the helm of the UNFCCC, Copenhagen, often referred to as "Hopenhagen" following a so-named international advertising campaign launched by the UN in support of a comprehensive treaty to emerge from COP15, instead descended into chaos, turning "the most significant effort ever by a Danish government to position itself as a global, political leader [...] into the biggest inter-national, diplomatic defeat for decades."[34] Such was the extent of the disorder that when then US Secretary of State Hillary Clinton welcomed President Barack Obama at the last day of negotiations, she apologet-ically declared it "the worst meeting I've been to since the eighth-grade student council."[35]

The result came not for a lack of trying. The Danish hosts had assiduously prepared the summit starting in 2007, even creating a new Ministry of Climate and Energy, the first of its kind anywhere in the world, to prepare for the COP. However, by 2009, the earlier window of opportunity had begun to close as climate change fell off government agendas in the wake of the global financial crisis and concerns over costs of effectively mitigating and adapting to changes in the global climate replaced the promise and optimism of earlier years. Other reasons prominently helped sink the conference, too, including a leaked draft text prepared by the hosts containing no binding requirements for the United States and Western powers to shoulder a greater burden than developing countries.[36]

In the event, the UNFCCC Secretariat and de Boer were unable to contain the text's impact and dispel the resulting suspicion and mounting distrust among conference delegates. Further, the new US administration led by President Barack Obama was not yet in a position to put any concrete actions on the table to suggest more aggressive domestic moves toward a low-carbon future necessary to underpin a meaningful treaty. What did emerge was a smallest common denominator compromise document, the Copenhagen Accord, cobbled together by the United States and the BASIC group of countries (Brazil, South Africa, India, and China) in "a last-minute deal in a back room as though the nine months of talks leading up to this summit, and the Bali Action Plan to which they had all committed two years previously, did not exist."[37] While not a complete catastrophe – COP15 established green growth as a viable approach to address climate change mitigation in the wake of the global financial crisis and introduced the idea of the Green Climate Fund (GCF) which would later bear fruit at COP16 in Cancun – it fell far short of expectations, not least those of the UNFCCC's Executive Secretary himself. The following June, a disheartened de Boer resigned his post.

Straight talk

Four years on from the disappointment of Copenhagen, UNFCCC Executive Secretary Christiana Figueres was ready to throw caution to the wind. A former Costa Rican anthropologist, public official, and diplomat with a wealth of international experience including in negotiations under the UNFCCC and the CDM, Figueres was brought in by Secretary General Ban to replace the determined yet ultimately hapless de Boer. Ban initially gave her space to put her own stamp on the process, taking a step back from his direct involvement in negotiations to focus on the sustainable development agenda and efforts to replace the Millennium Development Goals.[38] Figueres kept on core officials, including as her deputy Richard Kinley, a Canadian diplomat who had joined the bureaucracy when it was still an ad hoc secretariat under Cutajar in 1993, and set about galvanizing a dispirited staff still reeling from the aftermath of COP15.[39] From the start of her tenure, she followed the more activist path laid out by her predecessor. But while both Figueres and de Boer "took a more political, public line and used their platform to reach out to the world,"[40] it was under Figueres that the Secretariat began to project itself more than ever before.

COP19, held in Warsaw at the end of 2013, proved a case in point. The coal-friendly Polish host government decided to host the World Coal Association's International Coal and Climate Summit to coincide with the second week of UNFCCC negotiations, threatening to withdraw attention and impact from the climate convention process. Figueres, "convinced that the coal industry needed straight talk and should be part of the solution if it was to play any role at all," asked to speak at the summit, venturing into the lion's den to address coal company CEOs directly.[41] Environmental NGOs were opposed to the move and their protest outside the summit venue forced Figueres to enter through the back door.[42] But when the moment came, the Executive Secretary did not mince words. Reminding industry representatives of the need to stay within the global carbon budget if targets were to be achieved, she was

> clear from the outset that my joining you today is neither a tacit approval of coal use, nor is it a call for the immediate disappearance of coal. But I am here to say that coal must change rapidly and dramatically for everyone's sake.

Figueres insisted that

> it is abundantly clear that further capital expenditures on coal can only go ahead if they are compatible with the 2 degree Celsius limit. [...] Some major oil, gas and energy technology companies are already investing in renewables, and I urge those of you who have not yet started to do this to join them. By diversifying your portfolio beyond coal, you too can produce clean energy that reduces pollution, enhances public health, increases energy security, and creates new jobs.[43]

No other Executive Secretary had ever spoken to a fossil industry gathering in this way and Figueres's intervention helped put the UNFCCC negotiations at COP19 back on track. More importantly, it was reflective not just of the style Figueres brought to the role but also of the progress that had been made since the UNFCCC's infant days in rhetorically connecting the core issues and directly addressing the energy problem at the heart of anthropogenic climate change. Figueres was not afraid to address fossil energy, and especially coal, as the main climate culprit "even if that ruffled some feathers."[44]

In the wake of COP15 in Copenhagen, a number of initiatives at various different levels began to complement the UNFCCC process. This included the global fossil fuel divestment movement that had come into its own in 2011/2012 following a report on unburnable carbon by the Carbon Tracker Initiative (CTI) and the 350.org campaign driven, first, by students on North American campuses and, later, all around the world.[45] Following COP19, Figueres, in a number of speeches and opinion articles, threw her weight behind the movement that aims to exert pressure on institutions to rid their investment portfolios of holdings in coal, oil, and gas companies with the overall goal of disincentivizing the further development of fossil fuel reserves. In an op-ed written for *The Guardian*, she called on faith groups and religious institutions to divest from fossil fuels as a morally right and financially prudent move.[46] Her efforts were boosted by Secretary General Ban and World Bank President Jim Kim who both backed the campaign in 2014, and the Governor of the Bank of England and Chairman of the G20's Financial Stability Board (FSB), Mark Carney, who, in a series of speeches, took up the issue of unburnable carbon and the associated risk of global financial markets carrying a carbon bubble. In March 2015, Figueres joined fellow alumni in calling on Swarthmore College to divest its endowment funds from fossil fuels. In a letter to the School's management and students, Figueres argued that "the risk return analysis of fossil fuel investments has shifted in the past five years," threatening the value of coal, gas, and oil assets. Highlighting the value of transferring investment into low-carbon assets, she further argued that "incorporating sustainability and climate factors into investment decisions adds financial value."[47] In her dual framing of divestment as both necessary to avoid the risk of diminishing assets and an opportunity to benefit from the increasing value of low-carbon, renewable investments, Figueres was able to draw on growing support from the movement, including from large charitable organizations such as the Rockefeller Brothers Fund which had declared its intention to divest from fossil fuels. It added a further dimension to the economic case for action first spelled out in more detail during de Boer's tenure, broadening the set of frames available to the Secretariat when engaging with different stakeholders.

Despite her and her team's strong advocacy in support of a deep and speedy low-carbon transition, Figueres knew that the UNFCCC

Secretariat would not be able to launch, let alone sustain, an all-out assault on fossil fuels. Instead, to make progress in negotiations the Secretariat "had to try and bring them on side given their critical role" in the global energy system and their influence and lobbying power in domestic capitals.[48] When Figueres began her work as Executive Secretary in 2010, the Secretariat had, in her words, the feel of a "trash bin" due to the poor state of the building's facilities as much as the discontentment among staff following COP15 in Copenhagen.[49] Figueres's tall task was to lift staff morale and put the process back on track. The latter included bringing an unprecedented number of stakeholders into the fold to avoid another diplomatic calamity in Paris. As a passionate campaigner, Figueres was "happy to talk to different audiences in a way that resonates with them."[50] Oil and gas majors were no exception.

And so, half a year after the Warsaw summit at a side event to the annual Carbon Expo conference in Barcelona, she argued it was "time to stop pointing the blaming finger at fossil fuel companies" and that "bringing them with us has more strength than demonizing them."[51] Specifically, Figueres pointed to the technical and engineering capacity of the oil and gas industry to develop innovative low-carbon solutions. The move was informed in part by the role afforded to carbon capture and storage (CCS) in the IPCCs AR5 as well as other authoritative projections on the world's ability to effectively reduce GHG emissions to the extent necessary contained in a range of energy outlooks, prominently those published by the IEA.[52] What they shared in common was the assumption that emissions reductions pathways in line with the 2°C stabilization goal (and even more so the 1.5°C ambition), first officially recognized in 2009 and later contained with legal effect in the Paris Agreement,[53] would be impossible by relying solely on renewables but would have to incorporate a decarbonization of the current and future consumption of fossil fuels. Through the acknowledgment of a role for industry in the years preceding COP21, the "oil majors recognized her [Figueres] as a legitimate voice" and "someone they could work with despite their obvious differences."[54]

But it was not only the Executive Secretaries' advocacy work that underwent a change, branching out to bring onboard as many stakeholders as possible and injecting more energy into the negotiation process. Along the rocky road from Copenhagen to Paris, observers began to also see a Secretariat figuring out exactly what it was supposed to be as an organization.[55] By the early 2010s, Figueres, Kinley and the Secretariat's leadership team pursued a vertical and horizontal broadening of efforts and a differentiation of organizational roles, in effect helping to further integrate heretofore more disparate governance architectures. Importantly, the Secretariat began to act as an "orchestrator" that sought to influence national governments through strategic interaction with various different non-state actors in the pursuit of shared policy goals, from cities to NGOs and multinational corporations, "demonstrating its capacity to be an autonomous actor in global climate policymaking."[56] The Secretariat developed and led new initiatives such as the externally funded Momentum for Change Initiative, showcasing concrete projects from around the world that seek to build a more climate-resilient, low-carbon future, and the Non-state Actor Zone for Climate Action (NAZCA) which provides a platform for commitments to reduce GHG emissions through the establishment of carbon prices, investment in renewables and energy efficiency measures, and the issuing of green bonds, among others, by a large number of non-state actors such as cities, companies, and investors and financial institutions.[57]

Semper porro

The Paris Agreement on climate change was a breakthrough for the international negotiation process. A total of 196 parties (195 countries plus the EU) agreed to

> holding the increase in the global average temperature to well below 2°C above pre-industrial levels and pursuing efforts to limit the temperature increase to 1.5°C above pre-industrial levels, recognizing that this would significantly reduce the risks and impacts of climate change.[58]

Unlike the more top-down Kyoto Protocol, the more bottom-up Paris Agreement grants country parties greater flexibility in achieving emissions reductions – through the so-called nationally determined contributions (NDCs). By creating a framework for voluntary pledges that can be compared and reviewed internationally and ratcheted up over time, the Paris Agreement sidesteps the kind of conflicts over top-down targets that bedeviled the Kyoto Protocol.[59]

The agreement further spells out the need for a clear roadmap to provide sufficient amounts of climate finance, sets a ratchet mechanism to increase ambition over time, and a Global Stocktake in 2023 to monitor implementation and progress. The willingness of the international community to stand by the Paris Agreement was evident in its fast-track ratification. In a historic development, the Agreement came into force within a year of its adoption, having passed the ratification threshold of 55 parties accounting for 55 percent of global GHG emissions.

Before negotiations in Paris began, President of the Conference Laurent Fabius set out the French "climate imperatives" in an op-ed published in *The New York Times.* Much of the editorial focused on the energy challenge at the heart of anthropogenic climate change. Fabius warned of "conflicts" over the supply of and access to fossil fuel resources, spanning a rhetorical arc from historical Franco-German struggles over coal deposits along the Rhine to modern day tensions between China and Japan over hydrocarbons in the East China Sea.[60] But while the framing of the climate change issue as fundamentally about energy, and the transition to renewable sources, was a clear acknowledgment of direct connections between the two issues, the Paris Agreement seems, on the face of it, to fare worse than the Kyoto Protocol with no mentions of energy and energy sources such as fossil fuels and renewables. However, unlike the first and ultimately unsuccessful attempt at international collective action on climate change, the Paris Agreement establishes an overall temperature target and calls for a rapid global peaking of GHG emissions neither one of which is achievable without fundamental changes to worldwide energy production and consumption. Further, unlike the Kyoto Protocol, the Paris Agreement needs to be understood not only through the lens of its main document, but rather the multitude of NDCs that give the international agreement actual meaning. Through them, a further integration of the energy and climate fields is realized. Although greatly varying in ambition, country submissions from Brazil to the EU to Indonesia have spelled out energy transitions and low-carbon development pathways as central to achieving the Paris Agreement's 2°C target. Reflecting on the success of COP21 in a speech in early 2016, UNFCCC Deputy Executive Secretary Richard Kinley emphasized that the NDCs and other elements of the Paris Agreement were bringing "the era of fossil fuels" to an end. "And real transformation of the energy sector is the will, and undertaking of all the world's governments, including fossil fuel exporters."[61] In this way, the Paris Agreement was consequential for other major international bureaucracies, too, including the IEA Secretariat and the World Bank. Both had lent their strong support to making COP21 a success and both would continue to do so in its aftermath, through sustained pro-climate advocacy as much as through an integration of the objectives of the climate convention into their own approaches and activities.

COP21 also produced further progress in the growing integration between climate and financial governance. Speaking at the negotiations in Paris, Mark Carney announced the creation of the Task Force on Climate-related Financial Disclosures (TCFD) under the roof of the FSB which includes G20 central banks and finance ministries but also IFIs such as the IMF and the World Bank.[62] Chaired by billionaire investor and former New York City mayor Michael Bloomberg, the TCFD would "make recommendations for consistent company disclosures that will help financial market participants understand their climate-related risks."[63] Critically, the TCFD's original recommendations and status reports published in the years since have all explicitly linked to the Paris Agreement and IPCC findings upon which the climate convention process is built.[64] Transposing the agreement into action on the ground requires governments to develop and enact ambitious NDCs but it also requires participants in the financial markets, including large development banks like the World Bank, commercial banks, and institutional investors to understand the role they need to play if the global economy is to decarbonize in line with the targets adopted by the international community at COP21.

Conclusion

The story of the UNFCCC Secretariat is one of transformation and adaptation to changing external circumstances. From the Kyoto Protocol to the disappointment of Copenhagen and then the Paris Agreement, the Secretariat guided the climate convention process through a turbulent two decades marked by an ever more urgent need for effective solutions to the climate emergency backed up by science as much as by the rise, in the climate negotiations, of a more assertive Global South in the face of US obstinacy. The international bureaucracy started off small, with few resources and staff and little organizational independence. Yet as climate change rose on the international policy agenda, the Secretariat grew, professionalized its work, and in due course began branching out beyond its initial environmental focus to incorporate economic issues and engage more actively with key players in the global energy and financial governance architectures. Energy and finance as the two keys to addressing the climate emergency are now ubiquitous to the Secretariat's internal and external work and communication. By integrating them into its operations and building broad coalitions, the Secretariat prepared the ground for an agreement to emerge from COP21, driving climate policy and governance integration elsewhere.

The moves pursued by the UNFCCC Secretariat came as the result of changes in the bureaucracy over time. The Secretariat has never enjoyed the kind of autonomy afforded to other international bureaucracies. This is in part because the climate emergency and responses to it are seen as "very politically charged and many governments are unwilling to let an international body interfere with the steps they take to reduce emissions."[65] Further, while it was created as an autonomous body under the UN Secretary General, the Secretariat was meant to facilitate the climate convention process and its conferences, not assume leadership of it. Consequently, for the early years of its existence, the Secretariat was caught in a "straitjacket" of "formal and informal rules" which "ruled out any proactive role or autonomous initiatives" and barred staff from "exercising any leadership vis-à-vis parties and from assuming a more independent role."[66]

Over time, however, the Secretariat moved from behaving more like a "technocratic bureaucracy," quietly operating in the background and supporting Parties to the Conference, toward a more self-confident and "activist bureaucracy" that helped shape the global debate on climate change, framed the issue as one of great urgency, and effectively called for more ambition in emissions reduction commitments. The Secretariat was able to do so, first, through a growth in expert staff – the number of officials grew from a handful in the earliest days under Cutajar to over 400 under Figueres and current Executive Secretary Patricia Espinosa, aided by a growing core budget.[67] This enabled the taking on of greater responsibilities following the Kyoto Protocol and the Paris Agreement, the growth of expertise within the Secretariat's ranks, and greater cooperation and information sharing with partner organizations through dedicated committees, technical expert groups, and other venues. Second, and more importantly, the Secretariat also owed the shifts in its approach to "leadership that was right for its time."[68] The more cautious and bureaucratic Cutajar carved out an early role for the Secretariat and grew the number of staff and budget considerably. This enabled the international bureaucracy to deal with growing responsibilities in the run-up to and wake of COP3 in Kyoto. The more activist de Boer and Figueres built on this enhanced capacity to reach out to, and form coalitions with, key players in the energy, business, and finance sectors without whom no post-Kyoto agreement could ever have hoped to be effective. As climate change reached new heights on domestic and international policy agendas in the 2000s, so the office of UNFCCC Executive Secretary afforded its holders with more visibility and a bully pulpit to raise awareness and frame the debate on climate change mitigation and adaptation in a language understood by key players in the energy and financial governance architectures. De Boer and, even more so, Figueres seized the opportunity to expand the reach and meaning of the climate convention beyond its original scientific and environmental focus to build both rhetorical and physical linkages between climate change and related areas. It is these efforts that supported processes of governance integration and contributed to securing an historic agreement at COP21 in Paris.

Notes

1 Author's interview with senior IHS climate and energy expert, 1 November 2016.
2 Although the IPCC is referred to by name only once in the Paris Agreement (in Article 13), the importance of IPCC reports to the negotiation process has increased markedly over time, with growing scientific consensus gradually forcing the hands of negotiating parties. The Paris Agreement specifically defines itself through "the need for an effective and progressive response to the urgent threat of climate change on the basis of the best available scientific knowledge," knowledge collated and reviewed by thousands of scientists and spelled out in the IPCCs AR5. An even greater number of scientists and experts gathered evidence incorporated in AR6 published in 2021 and 2022, in time for the UNFCCC's global stocktake in 2023, one of the key elements of the Paris Agreement.
3 Author's interview with former senior UNFCCC representative, 20 October 2016.
4 Author's interview with former senior UNFCCC representative, 25 January 2017.
5 Author's interview with former senior UNFCCC representative, 20 October 2016.
6 Ibid.
7 Author's interview with former senior UNFCCC representative, 25 January 2017.
8 Author's interview with former US negotiator to the UNFCCC, 14 February 2017.
9 Author's interview with former senior UNFCCC representative, 20 October 2016.
10 Author's interview with former senior UNFCCC representative, 25 January 2017.
11 Author's interview with former UK negotiator to the UNFCCC, 28 November 2016.
12 UNFCCC, Kyoto Protocol to the United Nations Framework Convention on Climate Change (Bonn: UNFCCC, 1998).
13 Sylvia I. Karlsson-Vinkhuyzen, "The UN, Energy and the Sustainable Development Goals," in *The Palgrave Handbook of the International Political Economy of Energy*, eds. Thijs van de Graaf et al. (London: Palgrave Macmillan, 2016), 115–138.

14 For the Pulitzer Prize-nominated investigative series into ExxonMobil's climate change activities see the InsideClimate News website at https:// insideclimatenews.org/content/Exxon-The-Road-Not-Taken.

15 Author's interview with senior IHS climate and energy expert, 1 November 2016.

16 Author's interview with former senior UNFCCC representative, 4 November 2016.

17 Bill McKibben, "Warning on Warming," *The New York Review of Books*, 15 March 2007.

18 Christoph Böhringer and Michael Finus, "The Kyoto Protocol: Success or Failure?" in *Climate Change Policy*, ed. Dieter Helm (Oxford: Oxford University Press, 2005), 253–281.

19 Author's interview with former US negotiator to the UNFCCC, 14 February 2017.

20 US Congress, Senate Resolution 98, 25 July 1997.

21 Julian Borger, "Bush Kills Global Warming Treaty," *The Guardian*, 29 March 2001.

22 Cheney personally rejected the scientific consensus on the reality of anthropogenic climate change. After the release of the IPCCs AR4 in 2007, which declared the scientific understanding of the reality of anthropogenic climate change as "unequivocal," he insisted that "we're going to see a big debate on it going forward, the extent to which it is part of a normal cycle versus the extent to which it is caused by man." ABC News, "EXCLUSIVE: Cheney on Global Warming," 23 February 2007.

23 Author's interview with former senior UNFCCC representative, 20 October 2016.

24 Author's interview with former UK negotiator to the UNFCCC, 8 November 2016.

25 Author's interview with former UNFCCC representative, 24 October 2016.

26 G8, *Gleneagles Plan of Action: Climate Change, Clean Energy, and Sustainable Development*, Gleneagles Summit, 6–8 July 2005, online at: https://assets.publishing.service.gov.uk/government/uploads/system/uploads/attachment_data/file/48584/gleneagles-planofaction.pdf.

27 Author's interview with former UK negotiator to the UNFCCC, 8 November 2016.

28 Author's interview with former UNFCCC representative, 24 October 2016.

29 Author's interview with former senior UNFCCC representative, 4 November 2016.

30 Per-Olof Busch, "The Climate Secretariat: Making a Living in a Straitjacket," in *Managers of Global Change: The Influence of International Environmental Bureaucracies*, eds. Frank Biermann and Bernd Siebenhüner (Cambridge: MIT Press, 2009), 245–264.

31 Nicholas H. Stern, *The Economics of Climate Change: the Stern Review* (Cambridge: Cambridge University Press, 2006).

32 Author's interview with former senior UNFCCC representative, 4 November 2016.

33 Author's interview with former senior UNFCCC representative, 20 October 2016.
34 Per Meilstrup, "The Runaway Summit: The Background Story of the Danish Presidency of COP15, the UN Climate Change Conference," *Danish Foreign Policy Yearbook* (2010), 113–135.
35 Mark Landler and Helene Cooper, "After a Bitter Campaign, Forging an Alliance," *The New York Times*, 19 March 2010.
36 Chris Carter et al., "When Science Meets Strategic Realpolitik: The Case of the Copenhagen UN Climate Change Summit," *Critical Perspectives on Accounting* 22, no. 7 (2011), 682–697.
37 Richard Black, "Why Did Copenhagen Fail to Deliver a Climate Deal?" *BBC News*, 22 December 2009.
38 Suzanne Goldenberg, "Ban Ki-moon Ends Hands-on Involvement in Climate Change Talks," *The Guardian*, 27 January 2011.
39 Janosch Delcker, "The Climate Revolutionary," *Politico*, 10 November 2015.
40 Author's interview with former senior UNFCCC representative, 20 October 2016.
41 Author's interview with former senior UNFCCC representative, 25 January 2017.
42 Clara Germani, "Climate Change Summitry's Force of Nature: Christiana Figueres," *The Christian Science Monitor*, 21 September 2014.
43 Christiana Figueres, *Speech to the World Coal Association International Coal and Climate Summit*, Warsaw, Poland, 18 November 2013.
44 Author's interview with former US negotiator to the UNFCCC, 10 November 2016.
45 Julie Ayling and Neil Gunningham, "Non-state Governance and Climate Policy: The Fossil Fuel Divestment Movement," *Climate Policy* 17, no. 2 (2015), 131–149.
46 Christiana Figueres, "Faith Leaders Need to Find Their Voice on Climate Change," *The Guardian*, 7 May 2014.
47 Suzanne Goldenberg, "UN Climate Chief Joins Alumni Calling on Swarthmore to Divest from Fossil Fuels," *The Guardian*, 24 March 2015.
48 Author's interview with senior expert connected to the UNFCCC Secretariat, 16 December 2016.
49 Janosch Delcker, "The Climate Revolutionary."
50 Author's interview with senior expert connected to the UNFCCC Secretariat, 16 December 2016.
51 Ed King, "Stop Demonising Oil and Gas Companies, says UN Climate Chief," *Climate Change News*, 26 May 2015.
52 Author's interview with senior expert connected to the UNFCCC Secretariat, 16 December 2016.
53 The 2009 Major Economies Forum that preceded COP15 in Copenhagen represented "the first time that a consensus had been reached between the main developed and developing countries regarding the 2°C target." See Yun Gao et al., "The 2 °C Global Temperature Target and the Evolution of the Long-Term Goal of Addressing Climate Change – From the

United Nations Framework Convention on Climate Change to the Paris Agreement," *Engineering* 3, no. 2 (2017), 272–278.

54 Author's interview with senior expert connected to the UNFCCC Secretariat, 16 December 2016.

55 Author's interview with senior IHS climate and energy expert, 1 November 2016.

56 Thomas Hickmann et al., "The United Nations Framework Convention on Climate Change Secretariat as an Orchestrator in Global Climate Policymaking," *International Review of Administrative Sciences* (2019), 1–18.

57 As of 2019, the total number of actions stood at 26,864 pledged by 18,043 actors. While NAZCA themes reach from land use to human settlements and industry, more than half of all actions (15,355) were proposed changes to energy production and consumption. See the UNFCCC NAZCA website at https://climateaction.unfccc.int/.

58 UNFCCC, *Paris Agreement* (Bonn, Germany: UNFCCC, 2015).

59 See Robert Falkner, "The Paris Agreement and the New Logic of International Climate Politics," *International Affairs* 92, no. 5 (2016), 1107–1125.

60 Laurent Fabius, "Our Climate Imperatives," *The New York Times*, 25 April 2015.

61 Richard Kinley, *Keynote Address: Climate Change and Energy after Paris*, Selwyn College, University of Cambridge, 22 January 2016.

62 The FSB first proposed the task force to the G20 in early November 2015. The G20 Finance Ministers and Central Bank Governors then requested the FSB to set up such a task force.

63 FSB, *FSB to Establish Task Force on Climate-related Financial Disclosures*, Press Release, 4 December 2015.

64 TCFD, *Final Report: Recommendations of the Task Force on Climate-related Financial Disclosures* (Basel, Switzerland: Financial Stability Board, 2017).

65 Author's interview with former UK negotiator to the UNFCCC, 18 May 2018.

66 Busch, "The Climate Secretariat: Making a Living in a Straitjacket," 245–264.

67 The core budget grew from $14.5mn for the 1996–1997 biennium to nearly $57mn for the 2018–2019 biennium. See https://unfccc.int/about-us/budget/financial-and-budgetary-matters.

68 Author's interview with former UK negotiator to the UNFCCC, 18 May 2018.

3 The IEA as an adaptive bureaucracy

Washington D.C., 14 July 1981: At a US Senate hearing, the IEA's first Executive Director Ulf Lantzke discusses the objectives of international energy policy, including the Agency's role in coordinating the activities of industrialized countries during times of oil supply disruptions. Lantzke argues that energy security requires long-term structural changes and a shift from an overdependence on oil to coal, natural gas, and nuclear power.[1] Renewable energy sources, even those more established at the time such as hydropower, and environmental concerns play no role in the Executive Director's prepared testimony.

Paris, 13 March 2020: IEA Executive Secretary Fatih Birol calls for an acceleration of the low-carbon energy transition away from fossil fuels and increased global ambition on climate change, arguing that the "historic opportunity" to green governments' recovery spending to dampen the economic impacts of the Covid-19 pandemic must not be missed.[2] A month later, Birol praises renewables as the only energy sources "holding up" during the pandemic-induced slump in electricity consumption, predicting that "the energy industry that emerges from this crisis will be significantly different from the one that came before."[3]

Energy security has been at the heart of the IEA's mission since its founding in 1974 and remains so today in a world fundamentally different from the Agency's infant days. Executive Secretary Lantzke's remarks to the US Senate Subcommittee on Energy, Nuclear Proliferation, and Government Processes, delivered seven years after the IEA's creation, do not demonstrate the Agency's failings as an international organization. Diversifying away from oil following the crises of the early and late 1970s was a prudent move for the oil consumer states of the Global North. Modern renewables such as wind and solar PV, meanwhile, did not yet play a role in the energy mixes of OECD countries and the IEA was not conceived to address the environmental impacts of energy use.

DOI: 10.4324/9781315661339-4

They do, however, illustrate the long way the IEA, often considered the world's leading authority on energy economics, has come during its nearly half century of existence. Criticized until recently as too favorable toward oil, coal, and gas and employing "a deliberate method to hedge ever increasing profits for the conventional energy sector,"[4] the Agency – regarded as "the world's foremost multilateral energy organization"[5] and "the world's gold standard for energy analysis"[6] – has restyled itself as a key player in the transition to a low-carbon future. As this chapter shows, the IEA broadened its issue portfolio over time to cover the full range of energy technologies, including renewables, and treat climate change not simply as one of many problems but as a key determinant of twenty-first-century energy policymaking. More than just addressing the core objectives of the climate convention process as part of its activities, the IEA Secretariat built partnerships with the UNFCCC, the IPCC, and other actors, in effect helping to draw the global climate and energy governance architectures closer together. Importantly, it also transformed itself into an authoritative voice advocating for the interrelated goals of shifting away from fossil fuels toward low-carbon alternatives and taking decisive action to address the climate emergency. These changes came neither easy nor overnight and reflect the international bureaucracy's continuing engagement with a fast-changing global policy environment.

On standby

Crude oil has been known to man for thousands of years, with early seeps discovered and exploited in ancient Mesopotamia and China. A modern industry did not develop until the second half of the nineteenth century, however, following drilling operations in the 1840s and 1850s in the Baku region along the Caspian Sea, in Eastern Europe, and at the world's first steam-powered well at Oil Creek, Pennsylvania in 1859. What followed was a dramatic increase in the production and consumption of an energy source that fueled unprecedented economic growth and prosperity, modern transportation to connect the farthest reaches of the globe, the rise of some of the world's most powerful multinational corporations, two world wars, countless lesser conflicts across the Global North and South, and the global climate emergency due to the GHG emissions created in its combustion. It was Winston Churchill who gambled that, before long, oil would make the world go round when he, in his role as First Lord of the Admiralty in the early 1910s, committed the Royal Navy to a shift from coal to oil as its main transportation fuel.[7] The US-led reconstruction of Western Europe after the end of the Second World War, including through the earliest

World Bank loans and the Marshall Plan thereafter, focused on crude oil and refined products as among the most strategically important resources, in the process reshaping usage patterns and redefining the relationship between developed oil consumers in what was later to become the Organisation for Economic Co-operation and Development (OECD) and the developing oil producers of the Middle East.[8] In the midst of a period of decolonization and nationalizations of oil sectors around the Persian Gulf, the Organization of the Petroleum Exporting Countries (OPEC) was founded by Iran, Iraq, Kuwait, Saudi Arabia, and Venezuela in 1960 in Baghdad, with the goal of coordinating and unifying the petroleum policies of the founding members and support their economic interests on an international stage vis-à-vis the consumers of the Global North.

The IEA is a child of these historical developments. Established in 1974 as an autonomous organization within the framework of the OECD, its founding was a direct response to the 1973 oil crisis that had come as the consequence of both a reduction in oil production and an oil embargo imposed by OPEC members in answer to the Yom Kippur War fought between Israel and a coalition of Arab states led by Egypt and Syria. OECD countries, the world's major oil consumers at the time, were hit hard by the resulting oil price shock but rather than reacting in a more coordinated fashion, they engaged in "competitive behaviours such as stockpiling and hoarding of oil reserves" which drove costs up even further.[9] The IEA's original role was, thus, to ensure the energy security of its members by preventing similarly problematic behavior in the face of future oil supply crises.[10] The IEA's founding document lists the aims of the Agency as helping OECD countries develop self-sufficiency in oil in case of an emergency, establish oil demand limitation measures, gather and share information on developments in the international oil market, coordinate effective collective long-term responses to oil import dependence, and build closer relations between oil-consuming and oil-producing countries.[11] Today, the currently 29 member states are required to hold sufficient oil reserves to maintain consumption for at least 90 days without further oil imports. In the years since the IEA's founding, oil from this strategic reserve has been released three times: at the outset of the 1991 Persian Gulf War; after Hurricane Katrina had wreaked havoc in the Gulf of Mexico, destroying critical oil production infrastructure in the process; and in 2011 in response to the Libyan Civil War.

With its strong original focus on emergency measures to be taken during oil supply crises and no major role for the bureaucracy with regard to coordinating member state energy policies more widely, some labeled the IEA as merely an "insurance regime"[12] and a "standby organization."[13] However, although not part of its formal mandate, the IEA began to consider the impact of energy production and consumption on the environment toward the end of the Cold War, establishing initial links between energy and environmental policy. The first Ministerial Statement and Conclusions on Energy and the Environment was released by the IEA on 9 July 1985. Although the wording of this statement was quite general, it nevertheless included agreement on using energy more efficiently, combusting coal in an environmentally acceptable way, increasing the use of natural gas, and promoting renewable sources of energy. Importantly, the statement called on governments to give "due weight to environmental considerations in formulating their energy policies" and take necessary measures in close consultation with the OECD and IEA.[14] This opened the door for the Secretariat to play a greater role in its capacity as an advisor on wider energy policy matters, not least concerning the environmental implications of continued fossil fuel use.

Environmental issues were kept on the agenda thereafter, but the IEA betrayed little optimism at the prospects of effective climate change mitigation and transition toward a lower-carbon future. In 1990, the year the IPCC published its First Assessment Report, Executive Director Helga Steeg conceded that "for the whole world economy to achieve CO_2 stabilisation within two decades would require massive and probably impossible efforts."[15] However, the Agency did begin cooperation with the IPCC through the IPCC/OECD/IEA GHG inventory program and this work has intensified in recent years. In 1993, the Ministerial Declaration and Recommendation on Energy and the Environment urged ratification of the UNFCCC as one of the outcomes of the Rio Earth Summit the year prior and requested the IEA to assess the potential benefits of joint GHG emissions reduction activities with non-member developing countries.[16] The establishment of the Climate Change Expert Group (CCXG) the same year, hosted jointly with the OECD, would, over time, enable closer engagement with the climate convention process. In 1993, the IEA also officially expanded its scope beyond the primary objective of ensuring energy security to include as further goals economic development and environmental protection.[17]

A balancing act

Passage of the Kyoto Protocol at COP3 in 1997 brought climate change into sharper international focus. While fossil fuels received no direct mention in the Protocol, it was clear that their combustion lay at the heart of the climate change conundrum. The energy sector was not yet part of the solution, however, with fossil fuel consumption and associated GHG emissions continuing to grow largely unabated. The IEA Secretariat's own World Energy Outlook (WEO) – the international bureaucracy's annual flagship publication prepared and published under the direction of the Chief Economist – projected an increase in energy consumption of 65 percent and CO_2 emissions by 70 percent until 2020 in a business-as-usual scenario, with 95 percent of the additional energy consumption coming from coal, oil, and gas.[18] The IEA supported the UNFCCC process but occupied a more middling position on the treaty, focusing on an analysis of energy policy and emphasizing market-based mechanisms such as emissions trading spelled out in the Protocol's flexibility mechanisms, rather than offering the Protocol its full-throated support. It also did not make a concerted push for renewable energy sources or energy efficiency that were at the time seen to play only a complementary role in any climate change mitigation strategy.[19]

In a 1998 publication entitled "Benign Energy?", the IEA questioned the feasibility of a large-scale transition toward renewable energy sources. Renewables should be a factor, but their wide-scale application was held back by a number of environmental problems such as habitat loss from large hydropower projects and undesirable visual impacts of wind farms.[20] The following year, the IEA's Executive Director, Robert Priddle reiterated the publication's views in an opinion article published in the Bulletin of the International Atomic Energy Agency in which he questioned "the glib assumption that renewables are all good for the environment and fossil fuels all bad."[21] This view remained as the IEA position for years to come. In a speech to COP7 in Marrakech in 2001, Priddle emphasized that the environmental impacts of energy use, including climate change, needed to be approached with a view to ensuring the security of energy supply, which was high on the international agenda following a global economic downturn and the 9/11 terror attacks. Renewables could contribute to achieving energy security "but no fuel or technology can be excluded. Carbon-intensive energy forms may also become environmentally-benign through technologies such as carbon sequestration. Advances in nuclear technology could resolve both safety fears and the dilemma of waste disposal."[22] The Kyoto Protocol, meanwhile, only received a passing mention from the Executive Secretary in his address to a conference that was key to finalizing most of the Protocol's operational details and moved parties toward ratification of the world's first climate treaty.

This more careful positioning may at least in part be explained by the US stance on the climate change mitigation more generally, and the

Kyoto Protocol specifically. The IEA was created following a proposal by US Secretary of State Henry Kissinger who was driven by a desire to both institutionalize Western cooperation in the face of oil supply shocks and reduce US dependence on Middle Eastern oil.[23] Although an American has never led the IEA as Executive Director, the biggest and most powerful OECD economy and single largest contributor to the IEA budget sought to have its views represented by the Paris-based Secretariat.[24] In 2000, Jonathan Pershing, a US State Department official and later head of the IEA's Environment Division, stated his belief that a George W. Bush administration would address climate change and regulate GHG emissions.[25] But even members of Bush's own administration, including Environmental Protection Agency Administrator Christine Whitman, were surprised when, contrary to his statements on the campaign trail, Bush rejected the Kyoto Protocol early in his first term in a letter to Republican senators, with National Security Advisor Condoleezza Rice later declaring the treaty "dead" in a meeting with European ambassadors.[26] Instead of climate protection and low-carbon renewable energy sources, Bush's energy task force, chaired by Vice President Richard Cheney, presented a new national energy plan aimed at greater development of emissions-intensive oil and natural gas.

For the IEA, these developments required performing a "diplomatic balancing act" between the US position and that of European members in support of the Kyoto Protocol.[27] While the IEA's consensus-based decision-making in Governing Board meetings meant that all member states had a voice at the table, reducing the United States' ability to shape the course of the bureaucracy on environmental issues, climate change was not yet a dominant force in internal debates. Installed renewable energy capacity had grown significantly in the OECD world but wind and solar power were not yet cost-competitive with established energy sources. To boot, members did not consider the IEA to be an "environmental protection agency for energy."[28] It was still very much "an institution mainly concerned with fossil fuels."[29] This left room for the IEA and its Executive Director to pursue a more cautious approach.

An exchange on US funding for the IEA in early 2002 between Under Secretary for Energy, Science and Environment in the US Energy Department, Robert Card, and the Chairman of the Appropriations Subcommittee on Foreign Operations in the US House of Representatives, Republican Sonny Callahan, illustrates the position the IEA was seen to occupy during the early George W. Bush years. Callahan argued that "the IEA supports the Kyoto Protocol, while the Bush administration does not. Why is the Department providing funding for this Paris-based organization that supports the Kyoto

Protocol?" Card responded that while Priddle had made statements at COPs regarding the IEA's long engagement with climate change and support for GHG emissions reduction measures, the administration did not know of "any endorsement of the Kyoto Protocol." The IEA had published a number of studies on climate policies, CO_2 emissions trends in different energy sectors, and potential pathways for reducing them, but these were in support of "a Convention goal; not a Protocol goal."[30]

Despite continued US resistance, the Kyoto Protocol eventually came into effect in early 2005. Following the G8 summit at Gleneagles that same year, the IEA's work on climate change was to increase considerably.[31] Prime Minister Tony Blair wanted to use the British G8 Presidency to advocate greater climate ambition now that Kyoto was in force. Heading into the summit, the Bush administration resisted any mention of the Kyoto Protocol in the final documents and "was willing only to engage in additional scientific research on the issue."[32] However, following intense lobbying by Blair and the rallying effect of the London terrorist attacks which struck in the midst of the summit, Bush relented and signed the final communiqué that addressed Kyoto and described climate change as "a serious and long-term challenge" which G8 countries would seek to "slow and, as the science justifies, stop and then reverse."[33] The more general leaders' statement was followed by a Plan of Action on Climate Change, Clean Energy, and Sustainable Development that spelled out specific roles for the IEA ranging from reviewing the energy efficiency of buildings to assessing the GHG emissions profiles of coal-fired power stations. The Plan of Action was a confirmation at the highest level of outcomes produced at the IEA Ministerial Meeting earlier in the year that had also included requests for further research on carbon capture and storage (CCS). IEA Executive Director Claude Mandil welcomed the Plan of Action as "fully consistent with our mission to promote energy security, economic growth and a cleaner energy future through energy efficiency and technology cooperation."[34]

Identity crisis

In the following years, the IEA further expanded its work on gathering CO_2 emissions statistics, and built and maintained a database on emissions policies pursued by member states. At COP13 in Bali, the IEA's new Executive Director, Nobuo Tanaka, called for greater ambition in climate protection. Drawing on IEA research, Tanaka warned that unless countries took strong and decisive action, "we may be facing a 57% growth in CO_2 emissions by 2030. This energy scenario would commit the world to an increase in temperature of up to 6 degrees. We cannot let this happen." Solutions were to be found in "strong political direction from Parties" to the UNFCCC, a greater push for energy efficiency in line with the IEA's recommendations, and "a much broader deployment of the fully demonstrated low-carbon technologies, namely renewables and, subject to national policies, nuclear power."[35] However, while the IEA began to adjust its focus with regard to climate issues, such statements in support of renewables were too little too late for some who saw the Agency as "for the most part not qualified to represent the interests of renewable energy at the international level" because of its role in primarily advancing the cause of fossil fuels and nuclear power.[36]

IRENA was created in 2009 as a new intergovernmental organization out of a desire "to promote the widespread and increased adoption and use of renewable energy with a view to sustainable development," alleviate "problems of energy security and volatile energy prices," and reduce "greenhouse gas concentrations in the atmosphere, thereby contributing to the stabilisation of the climate system, and allowing for a sustainable, secure and gentle transit to a low carbon economy."[37] A number of IEA member countries, led by Germany and, to a lesser extent, Spain and Denmark, had pushed for its creation, dissatisfied with what they perceived to be an institutional capture of the Secretariat by fossil fuel and nuclear interests and, consequently, a downplaying of renewables and their potential role in the energy sector.[38] Since the creation of the IEA, the number of energy technologies had proliferated along with the number of energy policies enacted worldwide. Innovative policies and incentive mechanisms such as feed-in tariffs were driving the rapid uptake of wind and solar PV as the two most promising modern renewables. The founding of IRENA did not come as the result of a particular event but rather as the culmination of a long-term international engagement with renewables.[39] With the German energy transition, the so-called Energiewende in full swing, installed capacity for wind and solar PV growing fast across a number of European countries, and renewable power generation costs tumbling as a consequence, IRENA was, in the words of German MP Herrmann Scheer, who had lobbied for the creation of the organization since 1990, "an idea whose time has come."[40] Unlike the IEA that counts only OECD countries as its members, IRENA, with its headquarters in OPEC member state Abu Dhabi, was conceived as a universal organization, currently counting 161 members across the Global South and North.

The impending arrival of IRENA as a new member of the global energy governance landscape threw the IEA Secretariat into crisis mode. High-ranking IEA officials "responded jealously"[41] and attempted to prevent the creation of an organization "they regarded as an intruder on their turf."[42] It is undeniable that the IEA's earlier conservative projections for installed renewable capacity (for example, in wind energy) were later outstripped by actual developments; yet the IEA was hardly alone in its assessment. The US Department of Energy, the World Bank, and even the European Wind Energy Association – the European wind industry's own roof organization in Brussels – all underestimated actual installed capacity.[43] Yet, following a period of heightened activity in global climate diplomacy and the growth of wind and solar power widely regarded as a key to solving the climate puzzle, the IEA simply was not seen to have done enough to continue as the only major international organization entrusted with research, analysis, and, critically, advocacy on behalf of a renewable energy transition. The shock with which the IEA Secretariat greeted the birth of a new rival led to internal change and innovation, including the upgrade of the Agency's renewable energy unit to a division staffed with a larger number of full-time analysts.[44] This, in turn, built the ground for developing renewable energy into a significantly bigger part of its issue portfolio, through a more expansive gathering of statistics, more expert workshops in both member and partner countries, including major emerging economies such as China and India, and publication of a growing number of in-depth reports.

In 2011, the IEA established the Renewable Energy Industry Advisory Board to enhance links between the Agency and leading renewable energy industry stakeholders. IEA reports that acknowledge the rapid scaling up and increasing cost competitiveness of renewable energy sources such as onshore wind and solar photovoltaics began to be issued on a regular basis. The change toward a more positive outlook is illustrated by two solar technology reports published in late 2014 which together spelled out an ambitious vision for solar energy to become the world's largest electricity source by 2050 and contribute to a significant reduction in CO_2 emissions, the core goal of global climate governance.[45] These efforts added to the Agency's moves since the mid-2000s to expand its issue portfolio by gathering more statistics on electricity production, trade and consumption on a monthly basis for all OECD member countries, and publish reports on various aspects of electricity markets. Ahead of the launch of the IEA's 2014 Energy Technology Perspectives report, IEA Director Maria van der Hoeven argued that rather than oil, it is electricity that "is going to play a defining role in the first half of this century as the energy carrier that increasingly powers economic growth and development."[46] And it was in electricity generation that renewable energy sources showed their strongest growth.

The IEA also began to cooperate with its erstwhile rival. In early 2012, only three years after the creation of IRENA, the Secretariat signed an official partnership agreement, targeting the development and publication of the IEA/IRENA Global Renewable Energy Policies and Measures Database, collaboration in technology and innovation, and the sharing of renewable energy statistics. Commenting on the agreement, IRENA Director General Adnan Z. Amin declared the Agencies to be "natural partners in the global quest to increase the deployment of renewable energy."[47] Since then, the organizations have held joint workshops and published a number of energy technology briefs, for example, on solar photovoltaics and electricity storage. The technology briefs are published jointly by IRENA and the IEA's Energy Technology Systems Analysis Programme. In January 2015, IEA Executive Director van der Hoeven was a featured guest speaker at IRENA's Fifth Assembly alongside UNFCCC Executive Secretary Christiana Figueres. Two months later she praised the "good cooperation" between the IEA and IRENA at the Berlin Energy Transition Dialogue.[48] This seemingly more harmonious relationship between the two Agencies can also be explained by the organizational differences that have remained between them. Unlike IRENA, the IEA takes a more holistic view of the energy system, underpinned by its extensive data-gathering and analysis across the whole range of energy issues, whereas IRENA is an organization dedicated to promoting and advocating for renewables.[49] The differing mandates allow for a de-facto division of labor between the two organizations, even if this does not mean that rivalry between them has ceased altogether: at the IEA's 2015 Ministerial Meeting, Fatih Birol, newly promoted from IEA Chief Economist to Executive Director, presented his vision for the Agency to transform itself into a truly global organization, broaden its core mandate, cooperate more closely with emerging economies, and become a "hub for clean energy technologies and energy efficiency."[50]

Climate of change

The IEA's engagement with renewable energy sources more generally, and IRENA as a potential rival more specifically, is not the only change the international bureaucracy underwent on the road from COP15 in Copenhagen to COP21 in Paris and beyond. Contrary to its earlier image as too close to fossil fuels and not promoting specific GHG reduction targets, though generally supportive of climate protection, the Agency restyled itself as an influential global advocate on behalf of a low-carbon transition and aggressive climate change mitigation in line with the 2°C goal which had become synonymous with global climate stabilization. The IEA created an Environment and Climate Change Unit and since 2008, all World Energy Outlooks (WEO) – the IEA's annual flagship publication – have devoted substantial attention to climate change. The 2008 WEO focused on climate change alongside prospects for oil and gas production. Specifically, it argued that "preventing catastrophic and irreversible damage to the global climate ultimately requires a major decarbonization of the world energy sources" and that COP15 in Copenhagen provided a "vital opportunity" to establish a "framework for long-term cooperative action."[51] The 2010 WEO included a new scenario setting out a policy pathway consistent with limiting the increase in global average surface temperatures to 2°C above pre-industrial levels which assumed strong pledges and action up to and beyond 2020.[52]

Importantly, WEOs fall under the editorial authority of the IEA's Chief Economist and senior leadership which helps in part to explain the continued focus on climate change when the issue had slipped down the international policy agenda, and that of IEA member states, during the 2008–2009 global financial crisis and its aftermath. Moreover, as one IEA official pointed out, there is little editorial interference from member states with the organization's reports, although country representatives are usually invited to provide comments, input, and suggestions.[53] Thus, while the interests of member countries are important in determining the Agency's overall direction, a degree of autonomy and culture of self-determined research also give IEA officials a chance to pursue more independent agendas and, as in the case of climate change, help reinvigorate the debate.[54] The growing focus on climate change had direct implications for ministerial officials involved in global climate negotiations. Publication of the WEO increasingly became a "regular date in the diary for climate change negotiators" which provided "incredibly helpful ammunition used in narratives with different countries."[55]

WEO Special Reports have also highlighted clear linkages between the energy and climate policy fields. In 2013, the IEA published "Redrawing the Energy-Climate Map," which defined changes in the energy sector as the "key to limiting climate change" and proposed a number of energy policy solutions to "help keep the door open to the 2°C target through to 2020."[56] This was followed by the 2014 Energy, Climate Change and Environment Insight Report presented to delegates at COP20 in Lima. In June 2015, the Agency published a WEO Special Report on climate change ahead of COP21 in Paris, which, explicitly emphasizing energy use and climate change as inextricably linked, spelled out a number of key steps needed from an energy perspective to achieve success at the

UN climate talks and beyond.[57] A stakeholder meeting presenting the preliminary conclusions of the report in early March 2015 was attended, according to IEA officials, by all the key players in the global climate policy arena, reflecting the IEA's convening power and its newfound "weight in the climate debates."[58]

Since 2007, the IEA has directly supported global climate negotiations by organizing workshops, seminars, and side events at UNFCCC COPs, which have facilitated the sharing of information and have helped build a better understanding of technical issues amongst country delegations.[59] This increased involvement sat alongside a broadening of the process supported by the UNFCCC Secretariat to include not just representatives from environment and foreign ministries, but also finance and energy ministry officials. In September 2012, the IEA and the UNFCCC Secretariat signed a Memorandum of Understanding aimed at reinforcing "mutual efforts to promote clean energy and combat climate change."[60] Both organizations would engage in a closer exchange, with the UNFCCC taking responsibility for the overall governance framework of climate change mitigation and adaptation and the IEA contributing its experience in energy policy and statistics within this framework.

Specifically, cooperation has developed along five major lines. First, the IEA supports the UNFCCC Secretariat in its efforts to establish reliable GHG emissions inventories through an expert review of emissions data – a core function of the climate convention. The data provided to the Secretariat by country parties is compared with and verified through emissions data provided by the IEA.[61] Second, the IEA plays a role in the UNFCCC's expert review process in relation to climate policy measures adopted by country parties to the climate convention. As the bulk of GHG emissions is tied to energy use, the majority of climate policy is necessarily also energy policy. Third, the IEA has provided input to the technical examination process, specifically the UNFCCC's Workplan on Enhancing Mitigation Ambition (Decision 1/CP.17) with the goal of scaling up decarbonization efforts in the pre-2020 period. IEA experts have participated in the Workplan's Technical Expert Meetings on issues ranging from energy efficiency and renewable energy sources to carbon capture and storage. Fourth, the IEA is also involved in the UNFCCC's Technology Mechanism (Decision 1/CP.16) and its two components, the Technology Executive Committee and the Climate Technology Centre and Network, both of which aim to enhance technology development and North-South transfer. This cooperation happens through the multilateral, public-private Climate Technology Initiative (CTI), an Implementing Agreement under the IEA. For example, the CTI's

Private Financing Advisory Network seeks to help mobilize private capital in support of climate-friendly, clean energy businesses operating in developing countries. Finally, the CCXG has provided technical input into the UNFCCC process although it remains separate from the climate convention.[62] It convenes two meetings per year between government, private sector, and civil society representatives that are also attended by the UNFCCC, organizes side events at COPs and the annual Bonn climate change conferences, and regularly publishes policy papers on issues relevant to the climate negotiations.[63] From 2016 to 2018, the IEA also supported the Ad Hoc Working Group on the Paris Agreement tasked with developing its implementation guidelines.

In addition to these formal and publicly visible activities, much of the collaboration has also been of a more informal nature, built on routine interactions between IEA, UNFCCC, and OECD member country officials. For example, during negotiations in Paris in 2015, Paul Watkinson, Head of the Climate Negotiation Team in France's Ministry of Ecology, Sustainable Development and Energy and a key player in preparing COP 21, was "regularly in and out of the IEA" having built a "close relationship" with experts at the Agency.[64] The IEA has also involved UNFCCC officials in shaping the WEO Special Reports through meetings at IEA headquarters in Paris.[65] Over time, this closer cooperation bore fruit. Among negotiators there was a strong sense that the Agency had gotten "better at doing climate work," and had added a "clear climate lens" to its assessments.[66] At the same time, the IEA's impact was amplified because it was perceived as an "independent body" providing "heavy-hitting evidence" in support of rapid decarbonization rather than an environmental organization.[67]

Although the number of IEA staff working specifically on environmental issues is still quite small and large parts of the Agency "go on cheerfully without addressing climate change directly,"[68] major IEA units, such as those working on renewables, energy efficiency, energy technologies, electricity markets, carbon capture, and coal, all increasingly address climate change and GHG emissions as a routine part of their work. In part this represents the IEA's engagement with a dynamic global policy environment in which climate change considerations have become gradually more prominent. Another factor has been the international bureaucracy's relatively high staff turnover. Permanent IEA contracts are rare and the majority of staff turns over every five to six years, resulting in the average age of mid-level staff in the late 30s and early 40s. As one IEA official put it, the Agency may not always have pursued a conscious strategy of recruiting staff with knowledge of climate policy and an understanding of its connections with energy policymaking, but the inflow of younger experts who have been more exposed to these issues than their predecessors has nonetheless significantly expanded the IEA's in-house understanding and expertise.[69]

An energy revolution

The IEA's new role is also reflected in the external change advocacy pursued by leading figures within the bureaucracy, among them Executive Directors Maria van der Hoeven and Birol. While not natural proponents of a low-carbon, climate change agenda, both nonetheless realized that climate change was revolutionizing the energy sector and that the IEA had to position itself effectively if it wanted to remain at the forefront of the debate.[70] The consequence was language scarcely heard in the first four decades of the IEA's existence. Ahead of the 3rd Clean Energy Ministerial held in London in April 2012, van der Hoeven, a former Dutch Minister of Economic Affairs, warned that the world's

> addiction to fossil fuels grows stronger each year. Many clean energy technologies are available but they are not being deployed quickly enough to avert potentially disastrous consequences. [...] The current state of affairs is unacceptable precisely because we have a responsibility and a golden opportunity to act [on climate change].[71]

Commenting on the 18th Conference of the Parties to the climate convention in Doha, Qatar, van der Hoeven emphasized "the need to rapidly transition to a more secure, sustainable global energy system" in which "carbon emissions must be dramatically reduced."[72] Birol, a former statistician at the OPEC Secretariat, used similar language to position the IEA and its 2011 WEO in the climate change debate. Following the report's release in November 2011 shortly before the conference of the parties in Durban, he chided governments for their insufficient, non-legally binding emissions pledges, arguing that

> with current policies in place, global temperatures are set to increase 6 degrees Celsius, which has catastrophic implications. [...] If as of 2017 there is not a start of a major wave of new and clean investments, the door to 2 degrees will be closed.[73]

In 2013, van der Hoeven welcomed the release of the first part of the IPCCs AR5 by calling for greater ambition in transitioning to a low-carbon future. She emphasized that while scientific evidence grew more and more overwhelming with each passing year, "action to transform the way we produce and consume energy is slow. Energy accounts for two-thirds of global greenhouse gas emissions and as such its role is central in tackling climate change."[74] Yet the IEA did not just react to IPCC findings; it helped to actively shape them, too. The IPCC's AR1 released in 1990 already drew on IEA scenarios and reports but the engagement was still limited. By the time IPCC experts gathered to pull together AR5, the relationship had changed exponentially. All but one of the 16 chapters making up AR5's mitigation-focused third part published in 2014 drew heavily on IEA statistics and publications, with particular emphasis on the chapters addressing the energy sector, transport, industry, and buildings. While differences in approach have remained between the two organizations owing to their different mandates, the developments are reflective of an increasing similarity in the framing of climate change and energy and the interconnections between them by both the IPCC and the IEA Secretariat.[75]

Birol also helped position the Agency at the forefront of the campaign against fossil fuel subsidies, a critical step toward leveling the energy playing field for renewable energy sources trying to gain a foothold in the market. As early as 2009, the G20, based on data provided by the IEA, had called for a phasing out of such subsidies in the "medium term."[76] Speaking to *The Guardian* in early 2012, Birol emphasized that "energy markets can be thought of as suffering from appendicitis due to fossil fuel subsidies [...] undermining the competitiveness of renewables," adding that such subsidies "are a hand brake as we drive along the road to a sustainable energy future."[77] In 2014, he warned against both the growing global use of emissions-intensive coal and continued fossil fuel subsidies, arguing that "there is a need to change course in a dramatic way."[78] Making a similar case, van der Hoeven argued that the "growing use of coal globally is overshadowing progress in renewable energy deployment, and the emissions intensity of the electricity system has not changed in 20 years. [...] A radical change of course at the global level is long overdue" in order to achieve climate targets.[79] Presenting new IEA data in March 2015 which indicated a stalling of global GHG emissions from the energy sector in 2014 (albeit at a historically high level), Birol called climate change "the most important threat facing us today." He went on to argue that the new data would provide "much-needed momentum to negotiators preparing to forge a global climate deal in Paris in December."[80] Birol was unequivocal in establishing the connection between energy and climate change governance, arguing that "any climate agreement reached at COP21 must have the energy sector at its core or risk being judged a failure."[81]

This change advocacy has continued beyond COP21. Since 2016, the IEA has published its annual World Energy Investment reports which trace capital flows across the different energy sectors and project investment trends. The 2017 report concluded that global coal-fired power generation had peaked in the 2015–2017 period and IEA Chief Economist Laszlo Varro argued that there was no "investor appetite" for new emissions-intensive generating capacity.[82] Speaking in response to the IPCC's special report on the impacts of a rise in global average surface temperatures of 1.5°C above pre-industrial levels, Birol urged more renewables alongside coal plant closures to realize the Paris Agreement's ambition and avoid dangerous climate change.[83] Yet critics have challenged the IEA over precisely this issue. In a 2019 letter to the Agency, leading climate scientists and institutional investors challenged the IEA over its failure to include a realistic 1.5°C scenario in its WEO, arguing that not doing so provided cover for less ambitious GHG emissions reduction pathways.[84] The IEA has defended itself against these allegations by pointing to its alignment with the Paris Agreement whose central agreed objective is to limit temperature increases to below 2°C. More recently, the IEA has won the support of some of its erstwhile critics. Launching the Sustainable Recovery WEO Special Report in the midst of the Covid-19 crisis, Birol appealed for strong action to avoid a post-pandemic rebound in global emissions. The report put forward a blueprint for a climate-friendly economic recovery, developed together with the International Monetary Fund (IMF), which, with a strong focus on wind and solar PV at its heart, aimed to make 2019 the "definitive peak in global emissions."[85] Environmental NGOs such as Friends of the Earth and Greenpeace applauded the move, along with the Institutional Investor Group on Climate Change that represents major international banks, pension funds, and asset managers in a drive to mobilize sufficient capital for the low-carbon transition.[86]

Conclusion

Scarcely an issue in the infant days of global energy governance, climate change has since become one of the key determinants of a modern, twenty-first-century energy policy. Like its members and associated countries, the IEA Secretariat has had to respond and adapt to this development. As this chapter has revealed, substantial changes took place within the bureaucracy from the mid-1980s onward that, over time, transformed it into a key player on climate change issues, contributing to processes of governance integration. These changes have included a broadening of the Agency's research and analysis to incorporate the entire spectrum of energy issues, including energy efficiency, renewable energy sources, and climate policies. The widening of its portfolio has cemented the Agency's position as the leading international organization in the energy field: no other organization can claim to cover the full range of energy issues with as much authority. The IEA Secretariat's increased data collection and more positive outlook on renewables have helped to build a stronger case for a transition away from fossil fuels and have enabled greater integration within global energy governance through a closer cooperation with IRENA. If, following Van de Graaf, the creation of IRENA in 2009 was meant to lead to the kind of "radical departure from our current energy path"[87] seen as necessary to effectively address the climate emergency, then the IEA has shown its ability to be part of such a change.

The IEA has also contributed to governance integration through a closer cooperation with the climate convention process and the scientific review process under the IPCC. Its expert input has strengthened the work of the UNFCCC Secretariat, for example, by producing reliable GHG emissions inventories, feeding into IPCC ARs and other reports, and supporting the UNFCCC's expert policy review and technical examination processes. The IEA has been present at all COPs to the UNFCCC to date and has supported negotiations through forums such as the CCXG. Since 2007, the Agency has stepped up its involvement with the convention process by organizing a growing number of workshops, seminars, and side events at the annual COPs and mid-year negotiations at UNFCCC headquarters in Bonn. The closer cooperation between the IEA and the climate convention has turned the Agency into a partner upon whose technical and policy expertise the UNFCCC Secretariat has come to rely. In the run-up to and aftermath of negotiations for a new climate agreement at COP21 in Paris, observers saw the IEA as having become "an important voice in the climate space" and a "help in getting countries to implement and build on their NDCs."[88]

The moves toward greater governance integration are the result of changes within the international bureaucracy. First, the IEA Secretariat responded to ministerial decisions that set out core priorities and work plans. For example, both the 1993 Ministerial Declaration and Recommendation on Energy and the Environment and the 2005 Gleneagles Plan of Action requested the IEA to intensify its activities across a range of issues relevant to the climate convention process. But while the IEA's member states exert a degree of control over the Agency's general direction, not least through the power of the purse, the observed activities are not merely an agent's response to its principals' (member states') directives. Consensus-based decision-making in the Governing Board has reduced the influence of individual members and the Secretariat enjoys autonomy in the way in which it responds to requests. Asymmetry of information – the IEA Secretariat's unrivalled expertise across the full spectrum of energy issues puts it at an advantage over its members states – has further cemented this autonomy. While member and partner government officials regularly provide input to the IEA's work, there is little to no editorial interference over the Agency's reports, including the flagship WEO. The IEA's increasingly consequential role in integrating the energy and climate governance fields derives, therefore, not from a designed strategy or top-down plan, but has emerged through its efforts in pursuit of its energy-centric mandate in a complex and fast-changing global policy environment. That is, rather than a response to explicit demands from its member states, the IEA's role in integrating global energy and climate governance emerges through organizational change and adaptation impelled by today's global policy environment and novel ways in which it is exercising its organizational autonomy. And it is also a self-interested drive for continued relevance and survival, a motive international organizations and their bureaucracies share with the states that created them.[89] At the IEA's birth, energy security, the IEA's core mission, was analogous to the security of oil supply but since then, oil markets have become less contentious, the definition of energy security has broadened considerably, and environmental aspects of energy production and consumption have become vastly more important. As one IEA official put it, if the IEA ignored climate change as a key determinant of twenty-first-century energy policies and did not seek to play an active role in addressing the problem, it "would be abdicating itself and render itself irrelevant in the energy debate."[90]

Second, structural changes have significantly strengthened the role of climate change and low-carbon energy solutions in the Agency. In addition to the establishment of a unit on environment and climate change, energy efficiency, technologies, and electricity market units have all increasingly integrated climate change into their regular work. The salient context for these developments has been the rising prominence of climate change issues in domestic and international policy agendas, making them increasingly unavoidable for both the international bureaucracy's senior management and its constituent units. The direct challenge posed by the creation of IRENA also pushed the IEA to pursue change and innovation, not just through the upgrade of its unit on renewables, but in its outlook on the entire energy sector. In addition, the comparatively high staff turnover that has seen younger experts joining the IEA's various units has led to a gradual introduction of a better understanding of the interconnections between energy and climate policy.

Third, the IEA's senior leadership has played a key role in driving the organizational change underpinning the governance integration observed. It did so as much through changes to the IEA's internal structure as through external advocacy. The full-throated support for the Paris Agreement, fossil fuel subsidy reductions, and a swift transition to a low-carbon future exhibited by current Executive Director Fatih Birol mark the preliminary end of an evolutionary process of several decades. Earlier Executive Directors recognized climate change as an issue and helped broaden the Agency's portfolio but were more cautious in their approach to key outcomes of the climate convention process and their support for renewable energy sources. With wind and solar PV still in the initial stages of growth in the late 1990s and early 2000s, and the Kyoto Protocol marred by a number of problems, not least the non-participation of the United States as the world's largest economy and energy consumer, this position may be understandable. More recently, however, leading figures including Birol have increasingly emphasized the urgency of mitigating climate change and transitioning to low-carbon energy systems, advocating more forcefully on behalf of these interconnected agendas and aligning with the core objectives of the climate convention. They have focused attention and kept the issue on the international policy agenda, including at times when climate change slipped down the international policy agenda, such as the 2008–2009 financial crisis and the Covid-19 pandemic. The framing of climate change mitigation as urgent and requiring nothing short of an energy revolution by key IEA publications and the bureaucracy's leadership have helped draw the energy and climate governance architectures closer together.

Notes

1 US Congress, *International Energy Agency and Global Energy Security Matters: Hearing Before the Subcommittee on Energy, Nuclear Proliferation, and Government Processes of the Committee on Governmental Affairs*, Ninety-seventh Congress, First Session, 14 July 1981.
2 Chloé Farand, "Governments Have 'historic opportunity' to Accelerate Clean Energy Transition, IEA says," *Climate Change News*, 17 March 2020.
3 Jillian Ambrose, "Covid-19 Crisis Will Wipe Out Demand for Fossil Fuels, Says IEA," *The Guardian*, 30 April 2020.
4 James Murray, "IEA Accused of "Deliberately" Undermining Global Renewables Industry," *Business Green*, 12 January 2009.

5 Thijs van de Graaf, "The IEA, the New Energy Order and the Future of Global Energy Governance," in *Rising Powers and Multilateral Institutions*, eds. Dries Lesage and Thijs van de Graaf (Houndmills, UK: Palgrave Macmillan, 2015), 79–95.

6 Fiona Harvey, "World Has Six Months to Avert Climate Crisis, Says Energy Expert," *The Guardian*, 18 June 2020.

7 Daniel Yergin, *The Prize: The Epic Quest for Oil, Money & Power* (New York: Simon & Schuster, 2009).

8 See, e.g., David S. Painter, "The Marshall Plan and Oil," *Cold War History* 9, no. 2 (2009), 159–175.

9 Ann Florini and Benjamin K. Sovacool, "Who Governs Energy? The Challenges Facing Global Energy Governance," *Energy Policy* 37 (2009), 5239–5248.

10 A discussion of energy security is beyond the scope of this chapter. The IEA itself defines energy security as the uninterrupted availability of energy sources at affordable prices, although the environmental sustainability of said energy sources is now generally considered a key further aspect. For a good introduction to the conceptual debate, see, e.g., Christian Winzer, "Conceptualizing Energy Security," *Energy Policy* 46 (2012), 36–48.

11 OECD, *Decision of the Council Establishing an International Energy Agency of the Organisation* (Paris: OECD, 1974).

12 Robert Keohane, "The Demand for International Regimes," *International Organization* 36, no. 2 (1982), 325–355.

13 James Katz, "The International Energy Agency: Processes and Prospects in an Age of Energy Interdependence," *Studies in Comparative International Development* 16, no. 2 (1981), 67–85.

14 Richard Scott, *IEA: The First Twenty Years. The History of the International Energy Agency, 1974–1994, Volume Two: Major Policies and Actions* (Paris: IEA, 1994).

15 Sonja Boehmer-Christiansen, "Scientific Consensus and Climate Change: The Codification of a Global Research Agenda," *Energy & Environment* 4, no. 4 (1993), 362–407.

16 Scott, *IEA: The First Twenty Years. The History of the International Energy Agency, 1974–1994, Volume Two: Major Policies and Actions.*

17 Ibid.

18 IEA, *World Energy Outlook 1998* (Paris: IEA, 1998).

19 Author's interview with former IEA official, 20 March 2018.

20 IEA, *Benign Energy? The Environmental Implications of Renewables* (Paris: IEA, 1998).

21 Robert Priddle, "Energy & Sustainable Development," *Bulletin of the International Atomic Energy Agency*, 41, no. 1 (1999), 1–6.

22 Robert Priddle, *Energy Security and Climate Stability Must be Made Compatible, IEA Tells COP7*, Statement by the IEA Executive Director, 8 November 2001.

23 Marloes Beers, "The OECD Oil Committee and the International Search for Reinforced Energy-Consumer Cooperation, 1972–3," in *Oil Shock: The*

1973 Crisis and Its Economic Legacy, eds. Elisabetta Bini et al. (London: I.B. Tauris), 142–171.

24 In the early 2000s, the United States contributed 25 percent of the IEA's real budget. See Craig S. Bamberger, *IEA: The First Thirty Years. The History of the International Energy Agency, 1974–2004, Volume Four: Supplement to Volumes One, Two and Three* (Paris: IEA, 2004).

25 Dana R. Fisher, *National Governance and the Global Climate Change Regime* (New York: Rowman & Littlefield Publishers, 2004).

26 Julian Borger, "Bush Kills Global Warming Treaty," *The Guardian*, 29 March 2001.

27 Author's interview with former IEA official, 20 March 2018.

28 Scott, *IEA: The First Twenty Years. The History of the International Energy Agency, 1974–1994, Volume Two: Major Policies and Actions.*

29 Thijs van de Graaf, "Obsolete or Resurgent? The International Energy Agency in a Changing Global Landscape" *Energy Policy* 48 (2012), 233–241.

30 US Congress, *Hearings: Energy and Water Development Appropriations for 2003, Subcommittee on Energy and Water Development*, One-hundred-seventh Congress, Second Session, 6 March 2002.

31 Florini and Sovacool, "Who Governs Energy? The Challenges Facing Global Energy Governance," 5239–5248.

32 John J. Kirton and Ella Kokotsis, *The Global Governance of Climate Change: G7, G20 and UN Leadership* (Farnham, UK: Ashgate, 2015).

33 G8, *Gleneagles Plan of Action: Climate Change, Clean Energy, and Sustainable Development*, Gleneagles Summit, 6–8 July 2005.

34 IEA, *IEA Welcomes G8 Action Plan on Climate Change*, Press Release, 8 July 2005. www.iea.org/news/iea-welcomes-g8-action-plan-on-climate-change.

35 Nobuo Tanaka, *Statement to the 13th Conference of the Parties to the UNFCCC* (Paris: IEA, 2007). https://iea.blob.core.windows.net/assets/imports/events/10/Tanaka_bali.pdf.

36 Bernd Hirschl, "International Renewable Energy Policy – Between Marginalization and Initial Approaches," *Energy Policy* 37, no. 11 (2009), 4407–4416.

37 IRENA, *Statute of the International Renewable Energy Agency* (Bonn, Germany: IRENA, 2009).

38 Thijs van de Graaf and Dries Lesage, "The International Energy Agency after 35 Years: Reform Needs and Institutional Adaptability," *Review of International Organizations* 4, no. 3 (2009), 293–317.

39 Federico Esu and Francesco Sindico, "IRENA and IEA: Moving Together Towards a Sustainable Energy Future – Competition or Collaboration?" *Climate Law* 6, no. 3/4 (2016), 233–249.

40 Hermann Scheer, *The Time Has Come*, Speech delivered at the Founding Conference of the International Renewable Energy Agency, Bonn, Germany, 26 January 2009.

41 Maria Kottari and Panagiotis Roumeliotis, "Renewable Energy Governance Challenges Within a 'Puzzled' Institutional Map," in *Renewable Energy*

Governance, eds. Evanthie Michalena and Jeremy Maxwell Hills (London: Springer, 2013), 233–248.

42 Thijs Van de Graaf, "Fragmentation in Global Energy Governance: Explaining the Creation of IRENA," *Global Environmental Politics* 13, no. 3 (2013), 14–33.

43 REN21, *Renewables 2012 Global Status Report* (Paris: REN21 Secretariat, 2012).

44 van de Graaf, "Obsolete or Resurgent? The International Energy Agency in a changing global landscape," 233–241.

45 IEA, *Technology Roadmap: Solar Photovoltaic Energy* (Paris: IEA, 2014); IEA, *Technology Roadmap: Solar Thermal Electricity* (Paris: IEA, 2014).

46 IEA, *Taking on the Challenges of an Increasingly Electrified World*, Press Release, 12 May 2014. www.iea.org/newsroomandevents/pressreleases/2014/may/taking-on-the-challenges-of-an-increasingly-electrified-world-.html.

47 IRENA, *New IRENA-IEA Partnership Will Heighten Technology and Innovation Co-operation*, Press Release, 16 January 2012. www.irena.org/newsroom/pressreleases/2012/Jan/New-IRENA-IEA-partnership-will-heighten-technology-and-innovation-co-operation.

48 Maria van der Hoeven, *With Adnan Amin in #Berlin*, Tweet, 26 March 2015. https://twitter.com/VanderHoeven_M/status/581131254646628352.

49 Johannes Urpelainen and Thijs van de Graaf, "The International Renewable Energy Agency: A Success Story in Institutional Innovation?" *International Environmental Agreements: Politics, Law and Economics* 15, no. 2 (2015), 159–177.

50 IEA, *Energy Ministers Set Course for New Era at IEA*, News, 18 November 2015. www.iea.org/news/energy-ministers-set-course-for-new-era-at-iea.

51 IEA, *World Energy Outlook 2008* (Paris: IEA, 2008).

52 IEA, *World Energy Outlook 2010* (Paris: IEA, 2010).

53 Author's interview with former IEA official, 8 June 2015.

54 Author's interview with former IEA official, 12 March 2015.

55 Author's interview with former senior UK negotiator to the UNFCCC, 17 March 2015.

56 IEA, *Redrawing the Energy-Climate Map: World Energy Outlook Special Report* (Paris: IEA, 2013).

57 IEA, *Energy and Climate Change: World Energy Outlook Special Report* (Paris: IEA, 2015).

58 Author's interview with former IEA official, 12 March 2015.

59 The IEA has been present at COPs since the very first meeting in 1995 in Berlin but only began supporting negotiations more directly in Bali. At COP21 in Paris, the IEA hosted or co-hosted 21 different events.

60 IEA, *Collaboration between IEA, UNFCCC Will Lead to Improved Data and Analysis on Climate Issues*, Press Release, 26 September 2012. www.iea.org/newsroomandevents/news/2012/september/name,31627,en.html

61 This and following based on author's interview with UNFCCC official, 10 March 2015.

62 The CCXG began by advising government officials of industrialized OECD members, Annex I countries under the UNFCCC, on technical issues but over time expanded its role to include developing country officials, especially those representing the IEA's associated members like Brazil, China, India, and Indonesia.

63 There are several mechanisms of cooperation between the IEA and the UNFCCC, for example, through the Global Fuel Economy Initiative that is also tied in with the G20 and the UN's Post 2015 Development Agenda. The examples discussed here are intended to illustrate the more structured instances of engagement.

64 Author's interview with former IEA official, 12 March 2015.

65 Author's interview with UNFCCC official, 10 March 2015.

66 Author's interview with former senior UK negotiator to the UNFCCC, 17 March 2015.

67 Ibid.

68 Author's interview with former IEA official, 12 March 2015.

69 Author's interview with former IEA official, 8 June 2015.

70 Author's interview with former IEA official, 20 March 2018.

71 Fiona Harvey, "Governments Failing to Avert Catastrophic Climate Change, IEA Warns," *The Guardian*, 25 April 2012.

72 IEA, *Statement by International Energy Agency Executive Director on COP 18*, 3 December 2012, www.iea.org/newsroomandevents/news/2012/december/statement-by-international-energy-agency-executive-director-on-cop-18.html.

73 Sylvia Westall and Fredrik Dahl, "Oil Price Could Strangle Economic Recovery Hope, IEA," *Reuters*, 24 November 2011.

74 IEA, *IPCC Report Emphasises Need for Energy Sector Transformation*, News, 27 September 2013. www.iea.org/news/iea-ipcc-report-emphasises-need-for-energy-sector-transformation.

75 Johan Eriksson and Gunilla Reischl, "Worlds Apart, Worlds Together: Converging and Diverging Frames in Climate and Energy Governance," *Globalizations* 16, no. 1 (2019), 67–82.

76 Jeff Mason and Darren Ennis, "G20 Agrees on Phase-out of Fossil Fuel Subsidies," *Reuters*, 26 September 2009.

77 Duncan Clark, "Phasing Out Fossil Fuel Subsidies 'Could Provide Half of Global Carbon Target,'" *The Guardian*, 19 January 2012.

78 Karel Beckman, "Gradual Change Will Not Save Us," *Energy Post*, 3 June 2014.

79 Nathanael Massey, "Radical Change in Global Energy Long Overdue," *Climate Wire*, 13 March 2014.

80 IEA, *Global Energy-related Emissions of Carbon Dioxide Stalled in 2014*, Press Release, 13 March 2015. www.iea.org/newsroomandevents/news/2015/march/global-energy-related-emissions-of-carbon-dioxide-stalled-in-2014.html.

81 IEA, *IEA Sets Out Pillars of Success at COP21*, News, 15 June 2015. www.iea.org/news/iea-sets-out-pillars-for-success-at-cop21.

82 Simon Evans, "Seven Charts Show Why the IEA Thinks Coal Investment Has Already Peaked," *Carbon Brief*, 11 July 2017.

83 Adam Vaughan, "Energy Sector's Carbon Emissions to Grow for Second Year Running," *The Guardian*, 8 October 2018.

84 Leslie Hook and Anjli Raval, "IEA's Climate Models Criticised as Too Fossil-fuel Friendly," *The Financial Times*, 3 April 2019.

85 IEA, *Sustainable Recovery: World Energy Outlook Special Report* (Paris: IEA, 2020).

86 Harvey, "World Has Six Months to Avert Climate Crisis, Says energy expert."

87 Van de Graaf, "Fragmentation in Global Energy Governance: Explaining the Creation of IRENA," 14–33.

88 Author's interview with former senior UK negotiator to the UNFCCC, 12 June 2018.

89 Richard Collins and Nigel D. White, *International Organizations and the Idea of Autonomy: Institutional Independence in the International Legal Order* (London: Routledge, 2011).

90 Author's interview with former IEA official, 8 June 2015.

4   The World Bank's unlikely climate leadership

Washington D.C., 16 January 1951: The World Bank approves its first ever major loan to support a significant expansion of coal-fired power generation in the Global South. Thirty million dollars are lent to South Africa's Electricity Supply Commission to build seven new thermal power stations, upgrade existing installations, and add transmission lines and transformer capacity. With the Bank's help, South Africa builds a modern energy system run largely on domestically abundant coal. While it helps the country meet its growing energy needs it also, over time, ensures that South Africans hold the largest per capita CO_2 footprint on the continent, "competing" only with gas and oil-rich Libya for the dubious title.

Bali, 10 October 2018: At the joint World Bank and International Monetary Fund (IMF) Annual Meeting, World Bank President Jim Yong Kim confirms the Bank's "very firm decision" to discontinue its financial support for a planned 500 MW coal-fired power plant in lignite-rich Kosovo on the basis that "our bylaws [require us] to go with the lowest cost option and renewables have now come below the cost of coal."[1] Having committed itself to helping meet the Paris Agreement's goals and providing new record-breaking amounts of finance for climate change mitigation, the decision effectively ends more than six decades of direct World Bank lending to coal power projects around the world.

In the more than seven decades since its founding, the World Bank has provided financial support for countless infrastructure projects, many of them in the energy sector given energy's essential role for generating economic growth and development as prerequisites for achieving the Bank's twin goals of ending extreme poverty and promoting shared prosperity around the world. Through its support, the Bank has helped lift millions out of poverty; yet it also helped to enshrine high-carbon energy systems. Although the Bank reduced the number of loans it granted to emissions-intensive coal and hydrocarbon projects over time,

DOI: 10.4324/9781315661339-5

the last several years have nevertheless presented a step change in the international bureaucracy's activities in the energy and climate space. A shift away from fossil fuels did not come easy for the world's largest MDB operating in the developing countries of the Global South.[2] It was, rather, a gradual process driven by key players in the Bank who responded to a number of challenges and pushed for change, often against internal as well as external resistance. Following a brief overview of the Bank's genesis and early activities, this chapter traces the international bureaucracy's struggle to position itself with regard to climate change and the low-carbon energy transition in a fast-changing global policy environment. It pays particular attention to more recent developments, especially the periods of change activity surrounding the creation of the Global Environment Facility (GEF) and the Kyoto Protocol, and the years between the arrival of Jim Yong Kim as the new Bank President and COP25 in Madrid. While the early 2000s were marked by a number of structural changes internal to the international bureaucracy, the second period brought about much more externally visible change, increasingly aligning the Bank with international efforts to address the climate emergency and transition to a low-carbon future.

A "dirty" legacy

What would later become known as the World Bank was conceived at the 1944 Bretton Woods Conference as the International Bank for Reconstruction and Development (IBRD), then still a specialized agency of the UN tasked with, initially, providing loans for the rebuilding of war-torn Europe. The Bank officially opened for business in 1946 and made its first loan the following year, to France, for the acquisition of cars, trucks, and ships; petroleum; coal mining equipment; and raw materials.[3] Loans to the Netherlands, Luxembourg, and Denmark followed but with the Marshall Plan pouring significant investment into Western Europe from 1948 onward, the Bank's focus shifted to supporting economic development in the developing countries of the Global South. Given the need for sufficient amounts of energy provided reliably and at affordable prices – that is, energy security – to achieve economic growth and generate income necessary for recipient countries to repay their loans, much of the Bank's lending enabled or aided the construction of energy infrastructure: power plants, be they hydro-, petroleum-, or coal-fueled, and transmission and distribution grids. The Bank also developed a sustained interest in investments in hydrocarbon energy exploration and development following the oil crises of the

1970s.[4] Climate change and environmental sustainability had yet to appear on the policy agendas of the UN and international financial institutions as well as those of the developing countries receiving financial support. With cost-effective modern renewables still decades away and hydropower only suitable in specific locations, there was also often no viable technological alternative to coal- or petroleum-based thermal power, especially in countries with an abundant fossil resource base. Thus, from its early days the Bank, while hardly the only investor in such projects, helped enshrine a legacy of emissions-intensive infrastructure in power sectors across the developing world.

The International Development Association (IDA) which together with the IBRD makes up the World Bank was formed in 1960 to provide loans and grants to "poor countries that needed credit but could not afford it on market terms."[5] This new concessional financing mechanism signaled an expansion in the Bank's work, efforts that bore greater fruit under the transformational Presidency of Robert McNamara. In the McNamara years (1968–1981), the Bank's operations widened to cover more countries and more sectors beyond infrastructure; it greatly increased both the amount it loaned and the number of projects it took on and, consequently, more than tripled its staff from 1,600 to 5,700.[6] The Bank began to place greater emphasis on the needs of developing countries as opposed to only the interests of donors, and brought in as members such large borrowers as Egypt and China.[7] However, despite the widening of sectors to include agriculture, education, and others, the large increase in the amount of lending by both arms of the Bank – from roughly $1 billion in 1968 to $13 billion annually in 1981 – resulted in a further expansion of investments in emissions-intensive infrastructure and both on- and off-grid energy projects.[8] Environmental concerns, meanwhile, often took a backseat despite a number of forays under McNamara. In 1972, for example, the Bank positioned itself as a leader during and in the run-up to the United Nations Conference on the Human Environment, better known as the Stockholm Conference. When it seemed like developing countries would shun proceedings over what they perceived to be an attempt by the industrialized North to impose its views of environmental issues and, with it, shift undue economic burdens onto them, the Conference's Secretary General and later first UNEP Executive Director, Maurice Strong, called on the Bank to step in and convince developing countries of the importance of their participation and support. The successful intervention led to a compromise and a document, the Founex report, drafted inside the Bank that was to form the basis for the Stockholm Conference's Declaration, Principles, and its Action Plan.[9]

Internally, however, environmental issues continued to play only a comparatively marginal role in the bureaucracy's operations. By the 1970s and early 1980s, environmental loan conditions had been established but were often ignored, along with environmental policies in recipient countries.[10] Environmental NGOs had voiced little opposition to the Bank's operations throughout much of this time, in part due to a lack of inside knowledge of projects and the fact that the modern environmental movement had only just begun a decade earlier and was initially more concerned with addressing domestic environmental pollution, litigating against corporate non-compliers, and lobbying for environmental policy change.[11] But this changed with two high-profile cases that exposed the Bank to broad criticism from environmental NGOs and through media reports and, in turn, grabbed public and US legislative attention, culminating in a number of Congressional hearings.[12] In 1987, the Bank tried to quell public opposition that had mounted against the Polonoroeste road-building project in northwestern Brazil by establishing an Environment Department, regional environment divisions, and internal environmental assessment procedures. When these moves failed to satisfy the Bank's critics and the Sardar Sarovar Dam and Canal Project in India led to a new wave of antagonism over the resettlement problems it created, the Bank responded by deepening its environmental work and "mainstreaming the environment" across the bureaucracy's policies and administrative ranks from operational to managerial levels through the early to mid-1990s.[13]

Importantly, the organizational change that took place in the Bank in those years was the consequence of both external and internal dynamics. The Bank responded to pressure from NGO-led campaigns that threatened to become public relations disasters if left unaddressed and bowed to US Congressional oversight, but rather than simply reacting to and taking in these external influences like a homogenous actor, rifts within the Bank laid bare the bureaucracy's attempts to orient itself in a complex policy environment. Several of the Bank's technical experts warned against the potential loss of tropical rainforest cover and infringements on indigenous people's rights that could arise from the Polonoroeste project well before environmental NGOs made these the focal point of their attacks on the Bank's record, but their efforts were frustrated by project staff who wanted the Board to quickly approve loans and feared delays and increased costs.[14] The impact of the NGO-led campaigns gave support to those in the Bank who argued for making environmental issues a bigger concern and, over time, their ranks would grow, slowly introducing new thinking and helping to change the bureaucracy's approach from within. However, taking environmental aspects seriously when making lending decisions did not lead the Bank to terminate its investment in fossil fuel energy. The rise of climate change on domestic and international policy agendas would, in time, test the bureaucracy anew.

The Kyoto "catalyst"

From the early 1990s onward, global climate change gradually found its way into the Bank's activities as a response to both internal and external drivers. Externally, following a period of increasing political attention bestowed upon climate change in the late 1980s, the IPCC published its First Assessment Report in 1990 to much fanfare. Only two years later, in 1992, UNCED produced the UNFCCC as one of its main outcomes. With the climate convention process continuing its momentum toward adoption of the Kyoto Protocol at COP3 in 1997, forces within the Bank began to challenge established views held by entrenched parts of the bureaucracy that ranged from preferring to address "local environmental issues that mattered to projects on the ground rather than climate change as a global issue"[15] to "outright hostility to climate change [which] was seen as a constraint to the Bank's energy business."[16] A two-decade long struggle over integrating climate change into the Bank's operations was about to begin in earnest.

In 1991, the GEF was established within the Bank to invest in environmental protection measures and promote environmentally sustainable development. The GEF's three original implementing partners also included UNEP and the United Nations Development Programme (UNDP) but it was the Bank that drove the partnership and "controlled the way forward" for investments in developing countries.[17] Part of the Bank's efforts to answer its environmental critics, the GEF nevertheless came in for early opposition from recipient governments of the Global South who were instead looking for reparations for environmental damage already done, and NGOs who argued that in light of the Bank's track record, "giving [it] responsibility for global conservation was like putting a fox to guard chickens."[18] The restructuring of the GEF, completed in 1994, moved the facility outside the World Bank system, although the Bank remained highly influential as trustee of GEF funds. It further made the GEF the interim financial mechanism for the UNFCCC and the United Nations Convention on Biodiversity established at UNCED, and gave developing countries greater involvement in decision-making processes, steps that assuaged some of the concerns raised during the GEF's infant days. But while debates were raging externally, key officials within the Bank understood the GEF as a vehicle to drive organizational innovation, a "change agent" that could elevate environmental challenges, especially climate change, in a bureaucracy that had shown little sustained interest in such issues up to that point.[19] Ken Newcombe led GEF's first investment operations, assisted by Robert Watson, a leading atmospheric scientist who chaired the GEF's Scientific and Technical Advisory Panel and would later become Chair of the IPCC and the World Bank's Chief Scientist, and Charles Feinstein who, in his role as Team Leader for Climate Change in the Environment Department, advised the Bank on climate policies and helped assure the integrity of climate change-related projects under the GEF. By the end of its first decade the GEF had, through more than a thousand projects, leveraged around $4 billion in funding from the Global North to the South and former communist states, with annual funds three times as large as those available to UNEP.[20] Importantly, as a senior director, Newcombe had access to the new World Bank President, James Wolfensohn, and he used it to try and advance a climate agenda.[21]

During his tenure at the helm of the Bank, Wolfensohn was at least initially not particularly interested in climate change and sustainability. His focus on poverty eradication as the core mission of the Bank sat alongside moves to reform the bureaucracy, attracting more private capital to poor countries, ensuring the efficient and transparent use of resources and dealing with "the cancer of corruption."[22] When he did make the news on environmental issues, it was for his staunch support for projects such as the Chad-Cameroon oil pipeline to be built by a consortium led by ExxonMobil that was heavily criticized by international environmental and human rights groups for the damage it would do to ecologically fragile rainforests and indigenous peoples.[23] But over time, Wolfensohn took on climate issues as much in reaction to focusing events, the high stakes of the climate convention process and adoption of the Kyoto Protocol as to internal moves to popularize the issue and develop innovative new approaches to complement the Bank's lending and grant-making portfolio. The GEF was not the only such approach.[24] The run-up to Kyoto and the Kyoto Protocol itself acted as a catalyst in the creation of a number of carbon funds, opening up additional options for supporting climate change-related projects for the Bank. As early as February 1997, nearly a year before negotiations at COP3 were to begin, the Bank's Environment Department established an initiative to examine how a market-based carbon fund would be able to pool contributions from governments and private companies in support of a number of projects achieving GHG emissions reductions that would, in turn, be credited to individual investors on a proportional basis.[25] At the UN General Assembly's Special Session in June that year, Wolfensohn announced the Bank's intention to launch the fund if Parties to the UNFCCC were supportive, emphasizing that he saw the Bank's climate role "as providing every opportunity to developing countries to benefit from the huge investment" developed economies would have to make to reduce GHG emissions.[26] In the event, the Kyoto Protocol included three flexible market-based instruments – emissions trading, Joint Implementation (JI), and the Clean Development Mechanism (CDM) – which helped to pave the way for the Bank's proposal. The world's first carbon fund, the Prototype Carbon Fund (PCF), was launched in 2000 as a public-private partnership, following a consultation process involving potential contributors, host countries, the UNFCCC Secretariat, NGOs, and other stakeholders.[27] Led by Newcombe as fund manager, the PCF would mainly invest in renewable energy projects under the JI and CDM, with the Bank acting as host and "broker in helping to negotiate a price for the emission reductions that is reasonable for both buyers and sellers."[28]

The PCF was at once the logical next step in a process begun in the Bank ten years prior and a punctuation of the climate policy equilibrium. Later problems with CDM credits and uneven flows of climate and sustainable development co-benefits notwithstanding, the Bank embarked on a program of change that saw the establishment and growth of the Carbon Finance Unit (CFU) and, as of 2020, the development of more than a dozen carbon and climate funds and facilities – both multilateral and at individual donor country level – for which the Bank has acted as Trustee. The carbon finance work championed by the Bank operationalized the JI and the CDM and helped lead the way for the development of carbon markets around the world, including the European Union's Emissions Trading Scheme (EU ETS). But in 2000, investments in climate-friendly projects were still dwarfed by the Bank's more traditional hydrocarbon-oriented energy lending operations and the Bank had to speak to the range of issues spelled out in the UN Millennium Development Goals (MDGs) which did not explicitly address climate change. The Bank's energy and environment strategy paper "Fuel for Thought: An Environmental Strategy for the Energy Sector" published the same year sought to make the case that integrating environmental and climate change considerations into energy sector projects should "not be seen as a constraint but as a comparative advantage."[29] Although the paper, prepared by Andrew Steer of the Environment Directorate, Richard Stern of the Industry and Energy Department, Watson, and Feinstein among others, did not call for specific lending targets, it helped to influence the evolving thinking of energy sector managers in the Bank in two key ways. First, the realization began to take hold that "energy was not a sunset sector for the Bank and large amounts of new lending and investment were required."[30] Second, over the coming years many of those same managers "became increasingly convinced that renewables were hot property" and that climate finance could eventually eclipse IDA.[31]

Dr Watson, I presume?

Despite the boost the Kyoto Protocol provided, advocates of a shift away from fossil fuels toward renewables would continue to receive push back from the Bank's Board of Governors, including "China and India who were interested in continued coal funding to support economic development."[32] The conundrum was illustrated by the reception of the Bank's Extractive Industries Review. At the World Bank Group (WBG) annual meeting in June 2000, Wolfensohn had responded to continued criticism from the NGO community by promising to commission an independent review of the Bank's activities in support of oil, gas, and mining in developing countries. Published in 2003, the review found that there was still a role for the Bank in the sector but only if "interventions allow extractive industries to contribute to poverty alleviation through sustainable development." It also made a number of concrete recommendations, including the immediate ceasing of coal financing and the phasing out of oil investment by 2008.[33] The Bank Management's response the following year accepted the need for greater ambition but argued that "by staying engaged in oil and coal we can have an influential role in ensuring that the best environmental and social practices are followed and that the goal of sustainable poverty reduction is achieved."[34] Developing country members continued to request support for fossil fuel projects and the Bank would provide it, following evolving internal guidelines, for several more years. The numbers illustrate the uphill struggle for the Bank: while the PCF touted the likely reduction of 40–45 million tons of CO_2 equivalent emissions over a ten-year period in its 2004 Annual Report, the World Bank's own figures estimated an increase in these same emissions by around 300 million tons annually as the result of energy and mining projects in its portfolio.[35]

Adding to its difficulties, the US government under President George W. Bush was no friend of further integrating climate change into the Bank's operations and shifting away from fossil fuels. In early 2001, Bush withdrew President Bill Clinton's signature to the Kyoto Protocol and shortly thereafter unveiled a new national energy plan heavily skewed toward oil and gas interests. In a speech delivered shortly before the launch of the plan, Vice President Dick Cheney, chair of the task force that drew up the plan, claimed that "conservation may be a sign of personal virtue, but it is not a sufficient basis for a sound, comprehensive energy policy."[36] Taking its domestic agenda one step further, the White House went after leading climate voices in international forums and the Bank's Chief Scientist Robert Watson was high on the list. Watson, in his role as an outspoken Chair of the IPCC, had drawn the ire of ExxonMobil, the world's largest energy company and one of the main members of the President's Energy Task Force. In a 2001 memo to the White House acquired under the Freedom of Information Act, a lobbyist for the oil and gas major accused Watson and others of being "Clinton/Gore carry-overs with aggressive agendas" and asked if "Watson [can] be replaced now at the request of the U.S.?."[37] The Bush administration followed ExxonMobil's request and chose to nominate an alternative candidate, Rajendra Pachauri, rather than support Watson, leading to the first contested election in the IPCC's history. In the end, Pachauri won the vote "after a behind-the-scenes diplomatic campaign by the United States to persuade developing countries to vote against Dr Watson."[38]

Watson remained Chief Scientist at the Bank but here, too, the administration sought to take influence when former US Undersecretary of Defense and Bush-confidante Paul Wolfowitz was installed as Wolfensohn's successor. Joining the Bank in 2005, the same year the Kyoto Protocol came into effect following ratification by the Russian Parliament,[39] Wolfowitz and his team largely towed the Bush administration's line on climate change. This did not mean denying the issue in public – Wolfowitz addressed climate change and its impacts on poor communities in a number of speeches and welcomed the 2006 *Stern Review on the Economics of Climate Change* written by the World Bank's former Chief Economist and Senior Vice President Nicholas Stern – but de-emphasizing its importance, casting some doubt on its causes, and framing responses to the problem in the context of the economic development needs of the Global South.[40] On one heavily publicized occasion, the President and his deputy, Juan José Daboub, intervened to water down the science in a key Bank strategy paper and change the title to refer to "clean energy" instead of "climate change."[41] Watson and other scientists at the Bank pushed back, however, asserting their autonomy from interference, even if it meant indirectly challenging the most influential shareholder. In the end, they managed to extract sufficient concessions "to make the strategy paper credible."[42]

Ultimately, the Wolfowitz Presidency would prove but a short episode in the Bank's history, unable to meaningfully slow ongoing processes of governance integration. Forced to leave his post in 2007 in disgrace over a favoritism row, Wolfowitz made way for another Bush alumnus, former Undersecretary of State Robert Zoellick, who tackled climate change more head-on. In a speech marking the first 100 days as new Bank President, Zoellick committed to "step up our assistance to the international efforts to address climate change" and use both the Bank's Annual Meetings and COP13 in Bali "to outline a portfolio of ways the World Bank Group can help integrate the needs of development and low carbon growth."[43] The Bush administration's obstructive position on climate change led to its further isolation at the COP. Having blocked progress over language in the final text for the Bali roadmap, the US delegation was forced into an eleventh-hour U-turn, prompting one observer to argue that "the Earth's geopolitical axis had shifted slightly."[44] Zoellick, however, pushed in a different direction. Where the US delegation rejected financial help for poorer countries to combat climate change, he supported it. In the decisive final stages of negotiation, he argued for greater integration of development and climate agendas, reminding delegates that "climate change policies cannot be the frosting on the cake of development; they must be baked into the recipe of growth and social development."[45] He also launched a new carbon fund, the Forest Carbon Partnership Facility, prepared by the Bank with the aim of paying developing countries for avoided deforestation, and an informal discussion forum for finance and development ministers, the Bali breakfast meetings, which played an important role in UNFCCC Executive Secretary Yvo de Boer's efforts to link the climate convention process with international financial and economic development debates.

In 2008, following consultations with a range of stakeholders including recipients, donors, and UN agencies, the Bank established the Climate Investment Funds (CIFs) jointly with the AfDB, ADB, EBRD, and IDB as partners and implementing agencies. The goal was to provide scaled-up climate financing – in the form of grants, highly concessional loans, or risk mitigation instruments – to developing economies and speed up their transition to a low-carbon, climate-resilient future. While originally intended as an interim solution to provide urgently needed investment until the GCF was established and fully operational, the CIF, comprising a Clean Technology Fund (CTF) for scaling up low-carbon technologies and a Strategic Climate Fund (SCF) to help vulnerable countries adapt and build greater climate resilience, nevertheless quickly morphed into one of the world's biggest climate finance mechanisms and "the largest source of external concessional climate finance" for all of the multilateral development banks (MDBs) involved.[46] As with the GEF and the PCF before, the CIF represented an innovation in the landscape of global climate finance and an important tool to support the domestic adoption of low-carbon energy sources and, critically, cross-sectoral, collaborative approaches and policies necessary to effectively address the climate emergency.[47] The Bank's 2008 Strategic Framework for Development and Climate Change (SFDCC) created a footing to systematically pull together these and other approaches, including the issuance of one of the world's first green bonds. But the disappointment of COP15 in Copenhagen temporarily arrested the momentum in climate talks which had been building since Bali. At the COP, the Bank "was there to listen and support but stayed largely under the radar."[48] With expectations for the summit already through the roof and world leaders descending onto the Danish capital in their numbers, Zoellick "walked through the conference site almost unnoticed."[49]

The road to Paris

The long and rocky road from COP15 in Copenhagen to COP21 in Paris laid bare the Janus-faced position the Bank had maneuvered itself into by the turn of the decade: developing and growing innovative new approaches to carbon finance and embarking on an organizational change process to integrate climate change goals, while simultaneously continuing its support for emissions-intensive energy infrastructure responsible for creating the climate emergency. In 2008, the International Finance Corporation, the WBG's private sector arm that lends to businesses through financial intermediaries, had approved a $450 million loan to enable Tata Power Company to build a 4 GW supercritical coal-fired power plant in Gujarat, India. A year later a $180 million loan to India to renovate a fleet of aging, inefficient coal-fired power plants, co-financed by a $45 million GEF grant, followed. The Bank project leader defended the investment on climate change grounds, arguing that with effective modernization of coal-fired power stations, "India could be looking at emissions cuts anywhere between 10 to 13 million tonnes of CO2 equivalent each year."[50]

In 2010, the Bank's Board of Directors approved a $3.75 billion loan to South African power utility Eskom, the largest of South Africa's state-run enterprises and the biggest electricity producer on the African continent. The bulk of the loan, roughly $3 billion, would be used to help build the 4.8 GW Medupi coal-fired plant, the world's largest dry-cooled coal-fired power station upon completion in 2019. The rest of the money went toward piloting a utility-scale wind and a concentrated solar power plant, respectively, as well as energy efficiency measures.[51]

The project was in line with Bank policy: the SFDCC stated that in light of "the importance of coal for electricity generation in many developing countries, the WBG could support client countries in developing new coal power projects based on the most appropriate technology and analysis of alternatives."[52] Operational guidance published in early 2010 gave Bank officials the directions needed to decide on whether a coal project should go ahead.[53] Publicly, the Bank emphasized the need for large-scale investments in energy infrastructure following the 2007/2008 South African energy crisis which was marked by rolling blackouts due to Eskom's inability to meet increasing demand. Without the Bank's support, which came in addition to a loan granted by the AfDB, South Africa would "face hardship for the poor and limited economic growth," warned the Bank's Vice President for Africa.[54]

Behind the scenes, however, a battle raged over the present and future of the Bank's energy sector lending, described as "one of the most vicious in recent World Bank history."[55] With the support of environmental NGOs, the United States and the United Kingdom, the two largest donors, fervently opposed the loan on climate change grounds. Citing incompatibility with lending guidelines put in place by the Obama administration, the US Treasury argued that "without actions to offset the carbon emissions of the Medupi plant, the project is incompatible with the World Bank's strategy to help countries pursue economic growth and poverty reduction in ways that are environmentally sustainable."[56] But developing countries, including India and China, pushed back, claiming US attempts at influencing the Medupi decision "highlighted an unhealthy subservience of the decision-making processes in the bank to the dictates of one member country."[57] No action on US guidance should be taken "informally or formally, without an open, transparent and inclusive discussion and decision process."[58] The debate inside the Bank was no less vigorous, with climate and energy teams at loggerheads over whether to proceed.[59] But while answering the call for support from a major developing member country in crisis was a key argument, another line of reasoning helped to win over internal skeptics. If the Bank did not step up, leading figures warned, "the Chinese will. If we do it, at least we can do it in an environmentally friendly way."[60] In the end, the United States, the United Kingdom, Italy, the Netherlands, and Norway abstained from the Board vote, the traditional way to show dissent in the consensus-based body. But while the loan was granted, the ferocity of the battle quickened the Bank's transition away from coal. Further projects supported by the Bank would never again reach the scale of the Eskom deal. With climate scientists painting an ever more certain picture and the battle over an effective, low-carbon response to the climate emergency being fought both outside and inside the Bank, the writing was on the wall for the oldest fossil fuel.

Zoellick realized that more had to be done to address climate issues. Shortly after the Eskom deal had been approved, he appointed Andrew Steer as Special Envoy for Climate Change, a new position created at a level equal to that of Vice President within the Bank to coordinate all external climate change activities, oversee the CIF, and help integrate climate change into wider Bank operations.[61] Steer, a former UK civil servant and World Bank official, saw the Bank's role as "empowering innovation and to try and bring that sense of momentum and urgency into the international dialogue." The way to do so would be to produce "results at scale" via the CIF and, specifically, the CTF in leveraging total investment in renewable energy projects far greater than the amounts provided through Bank loans.[62] Steer also served on UN Secretary-General Ban Ki-moon's High-Level Group on Sustainable Energy for All (SEforALL), an initiative launched in 2011 to connect up core development objectives with action on climate change, the same task facing the Bank. The work begun under Steer acted as a proverbial bridge to the Jim Yong Kim Presidency which was marked by significantly greater activity on climate change than ever before.

Kim's arrival in 2012 gave a further shot in the arm to advocates of expanding the Bank's work in mobilizing carbon finance and other support for both climate change mitigation and adaptation projects. The former physician and Dartmouth President nominated for the post by US President Barack Obama came to the Bank with a "good understanding of the science and risks of climate change and its impacts on health and development" and made it clear from the start that the issue would be a "key priority" during his tenure.[63] Asked about his position on lignite coal projects in Kosovo at an event held at Brookings Institute three weeks into his new job, Kim spoke about the necessary "tradeoff between our need to keep the environment clean and poor countries' need for energy"; yet he also emphasized his views on climate change.

> I've been trained in science, and I think I'm actually the first one trained in science to be President of the World Bank, and so I have to tell you that the data that I'm seeing about changes that are happening today that we didn't think would happen for three or four years, the impact of a one-degree rise in the temperature of the oceans that we didn't think would happen until we got to two to three degrees' rise in the temperature of the oceans, this is extremely

disturbing to me, and I think we have to put the science of climate change in front of all of our member countries. And I guarantee you that I will do that.[64]

The following year, he announced new guidelines for the Bank's energy sector that reflected the objectives of the SEforALL initiative and would only extend support for coal projects "in exceptional circumstances" in which there were "no feasible alternatives to coal and a lack of financing for coal power."[65] While this was not yet a strict "no" to coal, it nevertheless represented a further step in the evolution of the Bank's position on the issue.

In November 2012, Kim launched a Bank-commissioned report by the Potsdam Institute for Climate Impact Research and Climate Analytics on the impacts and risks of a 4°C warmer world would hold and why such a potential future must be avoided. In light of the failure of COP15 to produce a meaningful treaty and global GHG emissions continuing to grow, such a scenario began to look like a distinct possibility. In the report foreword, Kim emphasized that the anthropogenic nature of climate change was "unequivocal," that the Bank committed itself to holding the rise in global average surface temperatures to below the widely accepted 2°C, and that he hoped that the dramatic impacts spelled out in the report – devastating floods and heat waves, food shortages, and malnutrition, among others – would "shock us into action."[66] At the report launch, Rachel Kyte, the Bank's Vice President for Sustainable Development and an ardent supporter of more stringent climate action, added that the "report reinforces the reality that today's climate volatility affects everything we do."[67] It was the first report published by the Bank, and one of only a small number of major international publications up to this point, which acknowledged in such stark terms the failures of international attempts to reduce GHG emissions and imagined a future in which climate change would not be effectively controlled. And it set up the internal and external changes the Bank would undergo leading up to COP21 in Paris.

Internally, Kim pushed a reform package to focus the Bank's efforts more closely on the twin goals of ending extreme poverty and promoting shared prosperity in a sustainable, climate-friendly way. This meant aligning the WBG's five branches for more effective operation, creating a new system of so-called Global Practices strongly focused on sustainable development, and prominently positioning climate change as one of five cross-cutting themes.[68] This climate "mainstreaming" made explicit the integration of climate change considerations into all programs and activities and across all levels of the Bank's operations and laid the ground for the Bank to sign up to the Principles to Mainstream Climate Action together with a number of other MDBs in Paris two years later.[69]

A concurrent staff shake-up saw two high-level departures and other executives reassigned to new roles.[70] Kim's first pick for the Bank's new executive line-up was Kyte who would continue as Vice President but with a beefed up portfolio to include Steer's role as Special Envoy for Climate Change after the latter left the Bank in 2012 to lead the World Resources Institute, and responsibility for integrating climate change into all the WBG's activities. In a *Financial Times* profile in early January 2014, Kyte called climate change the "ultimate curve ball" which threatened to undo development successes in the Global South and described Kim's vision of the Bank's new role on climate change as "standing on the top of mountains and shouting it out."[71] Kyte relished the opportunity to make a forceful case for climate action but her efforts "were much better received outside of the Bank than inside the Bank."[72] Internally, Kyte's more assertive personality and leadership style created some resistance across the Bank's practices. This was

> not because they didn't understand that climate change was not important but there were a lot of experts within the Bank who felt that climate change was maybe not as important to the governments they were dealing with as other issues such as poverty, health and education.

Persuading them would take some more diplomatic finetuning. Consequently, the Bank's climate and carbon finance teams engaged in a series of meetings "to convince other departments that climate change was a pressing issue which they and, in turn, partner governments needed to take more seriously."[73] Kyte herself admitted that "massive amounts of internal education" were needed when "you're going up against systems that have been put in place" and were hard to change.[74]

And change was coming, with the numbers revealing a considerable expansion in the Bank's climate practice over time. In 2014, the WBG counted nearly $12 billion, or close to 20 percent of its overall lending, as having climate co-benefits.[75] Shortly before COP21, Kim declared a target of 28 percent of all the Bank's lending to have climate co-benefits by 2020, equaling an increase in direct and leveraged co-financing of $29 billion at 2015 financing levels. The Bank would eventually exceed this target early, reaching 32 percent in 2018.[76]

While internal communication efforts faced some challenges, the Bank's external advocacy proved to be much more effective and "a new role started taking shape" for the international bureaucracy.[77] Kim and Kyte occupied a considerably more outspoken position and intensified cooperation with partners ahead of the historic year 2015 which was to produce both the Paris Agreement on Climate Change and the Sustainable Development Goals (SDGs) to replace the MDGs. One such position was Kim's support for the global divestment movement. Speaking at the World Economic Forum in Davos in early 2014, Kim made the case for a range of policies, including effective carbon pricing, a phase-out of fossil fuel subsidies, and forcing companies and financial institutions to disclose their exposure to climate risks. But he also called on leaders to "divest and tax that which we don't want, the carbon that threatens development gains over the last 20 years."[78] Like UNFCCC Executive Secretary Christiana Figueres before him, Kim's remarks gave credibility to a growing campaign that began on university campuses at a time when ExxonMobil considered it "a movement that is out of step with reality."[79] However, far from being out of touch, the movement continued to grow and have real world impact. In a 2018 report, Goldman Sachs noted fossil fuel divestment as a key driver in a 60 percent de-rating of coal producers since 2013.[80] And according to the campaigning website Gofossilfree.org, 1,237 institutions accounting for more than $14 trillion in value had announced divestment commitments by 2020.

Intensified cooperation also included engagement on climate change with the IMF and its Managing Director, Christine Lagarde. Sharing the stage for the first time at a climate event to open the 2013 Annual Meeting of the World Bank and the IMF, Kim and Lagarde made the case for a removal of fossil fuel subsidies, more action on developing carbon markets and pricing, and the need for greater investment in renewables and climate-smart agriculture.[81] Importantly, Kim emphasized the role both organizations could play in bringing "innovations that are happening in other parts of the world [and] are not always apparent to ministers of finance" to the table, helping governments integrate climate issues into their decision-making and understand "that they can create a better world for their grandchildren, but that it makes economic sense as well."[82] Later that year, Kim and Lagarde joined Figueres for the launch of the Green Climate Fund (GCF) in South Korea. The leaders' advocacy efforts continued at the World Bank/ IMF Spring Meeting in April 2014, this time with participation from UN Secretary General Ban Ki-moon. Kim, Lagarde, and Ban addressed the gathering of 46 finance ministers and senior officials, emphasizing the need for effective policies to reduce GHG emissions.[83] Ban asked attendees to come up with solutions to discuss at the UN Climate Summit held in September 2014 in New York City. The summit was intended to bring together public and private stakeholders and create momentum heading into COP20 in Lima and, ultimately, COP21 in Paris thatwhich was to produce a universal new climate change agreement.

Together with the Global Compact, the Bank had produced the statement "Putting a Price on Carbon" earlier in the year which called for greater collective action in "strengthening carbon pricing policies to redirect investment commensurate with the scale of the climate challenge."[84] Both also published a call for businesses to align with the "Business Leadership Criteria on Carbon Pricing" which were developed in cooperation with the UNFCCC Secretariat, UNEP, the Climate Group, the UN Foundation, and the Carbon Disclosure Project (CDP). Ban formally endorsed the calls for carbon pricing at the climate summit, along with 74 countries, 23 subnational jurisdictions, and over 1,000 businesses.[85] Before the summit, Kim, Kyte, and others had worked to assemble the broad coalition guided by the belief that "as you get more jurisdictions, you start building up the possibility of a bottom-up globally connected carbon market."[86] Ahead of COP21, the Bank had managed to position the need for effective carbon pricing as key to any eventual negotiation outcome.

Raising the bar

In the aftermath of COP21, the French Presidency and the UNFCCC Secretariat received praise for their deft handling of negotiations that culminated in the first viable agreement since the Kyoto Protocol nearly 20 years prior. Yet Bank insiders argued that it was the Bank that was instrumental in organizing the conference and, although it did not receive much public credit, ultimately helped to ensure its success. This included "making things happen and ensuring negotiators were aligned before even coming to Paris. The carbon finance team went around the world before the COP, they made sure everyone was looking at the same documents and negotiating beforehand."[87] The Bank, and especially its CFU, had every reason to be pleased with their efforts and the eventual outcome produced at COP21. Not only did the conference deliver a tangible core result: the Paris Agreement on Climate Change set an overall goal of keeping the rise in global average surface temperatures to below 2°C to be achieved through NDCs. Article 6 also specifically addressed carbon pricing through the creation of a voluntary carbon market in which it would be possible to use "internationally transferred mitigation outcomes to achieve nationally determined contributions."[88] In other words, such a system would make it possible to receive carbon credits for mitigation measures in one country, for example, a solar park, and count them toward achieving NDCs in another. This would allow for a trading system to be set up similar in nature to the CDM with which the Bank had already had significant experience, having run its carbon funds and facilities within the CDM framework for well over a decade. Problems with existing market arrangements, for example, the issue of double counting emissions reductions which arises when both the country selling carbon credits and the country buying them count emissions reductions toward their climate goals, made Article 6 the most contentious part of the negotiations, with agreement not forthcoming until the final day.[89] In early 2016, as countries around the world began signing the Agreement, the Bank and the IMF were busy "pressing governments to impose a price tag on planet-warming carbon dioxide emissions." John Roome, Senior Director for Climate Change insisted that "we've now got carbon pricing on the radar screen in a way it hasn't been before. We're moving from why to how."[90] But carbon trading has remained contentious in the years since Paris, with disagreement over the operationalization of Article 6 and other matters still unresolved at COP25 in Madrid, leading UN Secretary General António Guterres to lament that "the international community lost an important opportunity to show increased ambition on mitigation, adaptation and finance to tackle the climate crisis."[91] Time will tell whether effective rules to govern Article 6 can be agreed, but there is no doubt that the Bank has helped shape UNFCCC negotiations in fundamental ways.

Critically, together with other MDBs the Bank also pursued a strategy of aligning implementation of the Paris Agreement. At COP21, the ADB, AfDB, EBRD, EIB, IDB, and WBG pledged to jointly "increase our climate finance and to support the outcomes of the Paris conference through 2020."[92] More specifically, the MDBs promised to "support the US$100 billion a year commitment by 2020 for climate action in developing countries" as set out in the GCF, through a variety of actions, including the mainstreaming of investment principles, increasing access to concessional funding via the GEF, CIF, and GCF, provide greater technical and policy assistance, and use the MDBs' convening power to mobilize larger amounts of private capital.[93] These commitments were reiterated in the Bank's Climate Action Plan for 2016–2020 which spelled out more engagement on carbon pricing and the Group's efforts in "scaling up climate action, integrating climate change across its operations, and working more closely with others." The Bank called for "bold climate action" in implementing the goals of the Paris Agreement and the SDGs on the ground, including a commitment to screen all projects for climate risks and to account for the social cost of carbon in all its evaluations. It also pledged to step up its "global advocacy" with a focus "on selected issues where the WBG has an established voice – carbon pricing, mainstreaming climate action, and protecting the poor and vulnerable – with a focus on 'how' to deliver rather than 'why.'"[94] The plan recommitted the Bank to increasing the climate-related share of its lending portfolio to 28 percent from 21 percent annually. Having exceeded this amount in the 2017/2018 fiscal year already, the Bank pushed for greater ambition, announcing new climate lending targets for 2021–2025 in 2018 at COP24 in Katowice. The Bank now pushed for a doubling of its five-year investment projections to a total of $200 billion over the period. According to Kim, the new targets demonstrated "how seriously we are taking this issue" and that leadership by the Bank was intended to get the "global community to do the same."[95] Recognizing that climate change impacts were already materializing on the ground and threatening development gains across the Global South, and that adaptation had gained ever greater prominence in the climate convention process, the Bank earmarked a quarter of the investment ($50 billion) as direct adaptation finance.

And the Bank was also moving to come full circle in its support for fossil fuels. In 2019, campaigning group Urgewald published a report that accused the Bank of continuing to contribute to "higher profit margins for oil, gas, and coal operations" through both project finance and policy-based assistance.[96] The Bank's own figures indeed revealed a significantly higher investment in fossil fuel projects than renewables for the period 2014–2019. However, as the report acknowledged, much of the investment had legacy character – that is, the Bank's active portfolio included projects that were approved before the shift away from coal under President Kim and had not yet closed. Also included were loans for the clean-up of coal ash disposal facilities and coal mining sites, projects that may have financially benefitted the coal industry but hardly qualified as environmentally destructive. The criticism obscured that since the announcement of new coal guidelines for the Bank's energy sector, no new project finance to coal power stations had been granted and the Bank was openly advocating a shift away from coal. At a climate ministerial during the 2016 World Bank and IMF Annual Meeting, Kim singled out coal, especially projects earmarked by Asian countries, as the main stumbling block on the way toward a 2°C world as set out in the Paris Agreement. The Bank and its partners, Kim argued, "should all help to find renewable energy and energy efficiency solutions that allow them to phase out coal."[97] At the 2018 Annual Meeting, Kim then pulled the plug on the Bank's final potential coal-fired power plant project, the 500 MW power station in Kosovo he had been asked about at Brookings six years prior. Renewable alternatives were now cheaper and coal had become risky and undesirable from a climate change perspective, a point "the Bank strongly emphasized with its partners, leaving no doubt of its direction of travel."[98] Other fossil fuels were on the list, too. In 2017, at the One World Summit in Paris co-convened by the Bank, the UN and French President Emmanuel Macron, Kim and Guterres announced that "the World Bank Group will no longer finance upstream oil and gas, after 2019" and that only in "exceptional circumstances [will] consideration be given to financing upstream gas in the poorest countries where there is a clear benefit in terms of energy access for the poor and the project fits within the countries' Paris Agreement commitments."[99] In what must have seemed like an irony of history to an international bureaucracy used to being the frequent target of criticism over its many activities, environmental campaigners now praised the Bank as having "raised the bar for climate leadership."[100]

Conclusion

The story of the World Bank is that of an international bureaucracy gradually coming to terms with its indispensable role in driving the interconnected global climate and sustainable development agendas. Over the course of more than two decades, the Bank evolved to effectively position itself with regard to climate change and the low-carbon energy transition, with both internal and external forces seeking to leave their mark on its direction of travel. The Bank integrated key goals of the climate convention process into its operations, adjusted its lending guidelines and eventually ceased direct lending for coal and upstream oil and gas projects altogether, expanded its investment in climate mitigation and adaptation, grew the number of staff working on environmental and climate issues, created new internal units such as the CFU, forged closer links with other key players in the global climate governance architecture, and began using its platform to advocate for change.

The Janus-faced nature of the Bank during this period – continued support for emissions-intensive infrastructure projects at the very heart of the climate emergency and a concurrent development of innovative tools to finance low-carbon solutions – is a reflection of the bureaucracy's challenges in engaging with a complex and dynamic global policy environment marked by the urgent need to transition the global economy onto a climate-friendly footing to avoid the dangerous impacts of runaway climate change while simultaneously accommodating and supporting the rise of the Global South and its requirements for economic development and poverty alleviation. On the surface, the Bank has been a frequent target of critics for its seeming failure to effectively address the environmental impacts of a slew of energy and other economic development projects and for not moving away from fossil fuels fast enough. Yet, as this chapter has revealed, substantial changes took place within the bureaucracy from the late 1980s onward that, over time, transformed it into a leader on climate change issues and contributed to processes of governance integration.

In climate terms, the World Bank's current position represents a departure from the bureaucracy's earlier years. The creation of financial mechanisms and funds such as the CIF did not just enable projects through the provision of concessional financing; they leveraged several times their investment from private sources. This well-established de-risking power of MDBs, including the Bank as the biggest MDB operating in the Global South, lowers the cost of capital and crowds in more private sector investment as a key to mobilizing sufficient funds to help the world transition to a low-carbon future. The Bank's critics have pointed out that many of its initiatives did not achieve significant emissions reductions and the efficacy of market-based mechanisms championed by the Bank since creation of the PCF has rightly been the focus of much professional and academic debate. But the design and implementation of market-based mechanisms as key policy solutions have improved significantly since the early days of the CDM and the EU ETS and the continued growth in global GHG emissions is hardly the Bank's fault alone. To the contrary, its activities have contributed to making policy more effective and renewables and other low-carbon technologies more economic, which has, in turn, helped to make the case for the feasibility of the low-carbon transformation of the global economy. The Bank acts as a critical financier when no other sources of funding are available, is a facilitator and a policy advocate, but its influence only reaches so far. It is governments that need to enact and enforce the kind of legislation needed to comply with the Paris Agreement and the SDGs.

The moves toward greater governance integration came as the result of changes within the Bank. First, the Bank pushed the limits of its autonomy over time. Most of the Bank's capital comes from its borrowing on the international financial markets rather than member contributions which reduces its dependence on any one country. And yet as the largest and most influential shareholder and the only member with veto power over decisions taken by the Bank's Executive Board, the United States has always played an outsized role in Bank affairs. However, US attempts at frustrating the Bank's climate activities, or that of individual units within the bureaucracy as under Wolfowitz, were short-lived and at best managed to delay change. The Obama administration's attempts at halting Bank support for the Medupi power station failed, too, reflecting shifting global power dynamics and the Bank's attempts to effectively balance its role between the Global North and South. Second, structural and budget changes signifi- cantly strengthened the role of climate change and carbon finance in the Bank. These changes, including the establishment, capitalization, and growth of the GEF, the PCF, the CIF, as well as creation of the CFU and climate-focused roles in senior management, moved climate change from one of many issues to the heart of the Bank's operations. Climate and carbon finance only played a small role in the 1990s. Today, roughly a third of all the Bank's investments are climate related. These changes were as much a way for the Bank to answer its critics and respond to growing requests for financial and policy support from recipient members as they were a consequence of innovative solutions developed by key players on the inside who understood the reality of anthropogenic climate change and its growing impacts on the Bank's core mission.

Finally, the Bank's senior leadership played a critical role in driving the organizational change that underpinned the governance integration observed. They did so as much through internal reforms as through external advocacy and coalition-building. Zoellick recognized climate change as an issue much more than Wolfowitz and took steps to enable the Bank to address the issue in concert with its focus on combating poverty. But the Kim Presidency marked a step change in both internal and external activity which led one expert in the Bank to exclaim that as far as carbon finance is concerned, "the sky was the limit."[101] Kim and his core leadership team used their bully pulpit to help drive the debate and position the Bank and the financial solutions it had championed over a number of years as an integral part of the climate convention process. The proactive stance brought the Bank in closer alignment with the goals of the Paris Agreement. In its shift away from fossil fuel financ- ing, its ever-greater lending for both climate mitigation and adaptation measures, and its clear commitment to climate change as one of the five cross-cutting strategic priorities affecting all of its activities, the Bank of today is a much transformed bureaucracy. Although climate enthusiasm seems to have waned somewhat since Kim left the Bank in 2019, three years before the end of his second term, there is no indication that this is set to change significantly under current President David Malpass. Although installed by the Trump administration, whose defining if short-lived policy on climate change was to withdraw US participation in the Paris Agreement, Malpass has so far largely committed himself to the priorities set by his predecessor.[102]

Notes

1 Karl Mathiesen, "World Bank Dumps Kosovo Plant, Ending Support for Coal Worldwide," *Climate Change News*, 10 October 2018.
2 Judged by the volume of its annual borrowing and lending, the European Investment Bank (EIB) is the world's largest MDB. However, the EIB concentrates most of its operations within Europe while the large majority of projects (co-)financed or otherwise supported by the World Bank sits outside the OECD countries.
3 World Bank, *France Reconstruction Project: Loan 0001, Loan Agreement* (Washington, DC: World Bank, 1947).
4 Devesh Kapur et al., eds., *The World Bank: Its First Half Century*, Volume 1: History (Washington, DC: Brookings Institution Press, 1997).
5 Ibid.
6 Jochen Kraske et al., *Bankers with a Mission: The Presidents of the World Bank, 1946–91* (Oxford: Oxford University Press, 1997).
7 Following the oil crises in the early and late 1970s, the Bank also began to place greater conditionalities on borrower countries to adjust their economic policies in exchange for financial support, including in the energy sector. These structural adjustment loan programs would later become a source of sustained, heavy criticism of the Bank's operations for creating conditions at odds with the Bank's declared goals of poverty alleviation and more equitable economic development.
8 Infrastructure investment made up two-thirds of the Bank's lending portfolio in the 1950s and 1960s. While this share decreased to one third in the 1970s and 1980s, it still represents an increase in the absolute amount due to the dramatic overall increase in lending under President McNamara. See Robert H. Wade, "Greening the Bank: The Struggle for the Environment," in *The World Bank: Its First Half Century*, eds. Devesh Kapur et al. (Washington, DC: Brookings Institution Press, 1997), 611–733.
9 Wade, "Greening the Bank: The Struggle for the Environment," 611–733.
10 Bruce Rich, *Mortgaging the Earth: The World Bank, Environmental Impoverishment, and the Crisis of Development* (Washington, DC: Island Press, 1994).
11 Wade, "Greening the Bank: The Struggle for the Environment," 611–733.
12 See, e.g., Ian A. Bowles and Cyril F. Kormos, "Environmental Reform at the World Bank: The Role of the U.S. Congress," *Virginia Journal of International Law* 35, no.777 (1995), 813–819.
13 Wade, "Greening the Bank: The Struggle for the Environment," 611–733.
14 Robert H. Wade, "Boulevard to Broken Dreams, Part 1: the Polonoroeste Road Project in the Brazilian Amazon, and the World Bank's Environmental

and Indigenous Peoples' Norms," *Brazilian Journal of Political Economy* 36, no. 1 (2016), 214–230.

15 Author's interview with senior World Bank official, 20 June 2016.
16 Author's interview with senior World Bank official, 8 July 2016.
17 Author's interview with senior World Bank official, 14 June 2018.
18 Zoe Young, *A New Green Order?: The World Bank and the Politics of the Global Environment Facility* (London: Pluto Press, 2002).
19 Author's interview with former senior World Bank official, 27 June 2016.
20 Young, *A New Green Order?: The World Bank and the Politics of the Global Environment Facility*.
21 Author's interview with senior World Bank official, 20 June 2016.
22 James D. Wolfensohn, *People and Development: Annual Meetings Address*, World Bank presidential speech, Washington, DC, 1 October 1996.
23 Paul Brown, "World Bank pushes Chad pipeline," *The Guardian*, 11 October 1999. The project, planned and prepared since the mid-1990s with Wolfensohn's involvement, was approved in 2000. After Shell and Elf (now Total) left the original consortium, the Bank's continued support was critical in attracting further investors. The Bank canceled the agreement in 2008 after Chadian President Idris Déby had spent revenues on arms and a military expansion instead of poverty reduction, healthcare, and education measures as contractually agreed. At this point, however, the pipeline had long been built and the Bank's loans had been repaid. See Xan Rice, "World Bank Cancels Pipeline Deal with Chad after Revenues Misspent," *The Guardian*, 12 September 2008.
24 The Bank also launched a number of policy initiatives, including the Clean Coal Initiative aimed at encouraging policy reforms and the adoption of environmentally friendly technologies necessary for a "cleaner and more efficient use" of the most emissions-intensive fossil fuel. Peter van der Veen and Cynthia Wilson, "A New Initiative to Promote Clean Coal," *Finance and Development*, 34, no. 4 (1997), 36–39.
25 David Freestone, *The World Bank and Sustainable Development: Legal Essays* (Leiden, Netherlands: Martinus Nijhoff Publishers, 2013).
26 James D. Wolfensohn, *Towards Global Sustainability*, Remarks to the United Nations General Assembly Special Session on the Environment, New York, 25 June 1997.
27 Freestone, *The World Bank and Sustainable Development: Legal Essays.*
28 Johan Albrecht and Delphine François, "Voluntary Agreements with Emission Trading Options in Climate Policy," *Environmental Policy and Governance*, 11, no. 4 (2001), 185–196.
29 Author's interview with former senior World Bank official, 27 June 2016.
30 Author's interview with senior World Bank official, 20 June 2016.
31 Author's interview with senior World Bank official, 8 July 2016.
32 Author's interview with senior World Bank official, 20 June 2016.
33 Emil Salim, *Striking a Better Balance: The World Bank Group and Extractive Industries*, The Final Report of the Extractive Industries Review, Jakarta,

Indonesia   (2003).   https://lawweb.colorado.edu/profiles/syllabi/banks/
EIR%20vol1_eng.pdf.

34 World Bank, *Striking a Better Balance: The World Bank Group and Extractive Industries, The Final Report of the Extractive Industries Review – World Bank Group Management Response* (Washington, DC: World Bank, 2004).

35 Robert Marschinski and Steffen Behrle, "The World Bank: Making the Business Case for the Environment," in *Managers of Global Change: The Influence of International Environmental Bureaucracies*, eds. Frank Biermann and Bernd Siebenhüner (Cambridge: MIT Press, 2009), 101–142.

36 James Carney and John F. Dickerson, "The Rocky Rollout of Cheney's Energy Plan," *TIME Magazine*, 19 May 2001.

37 Randy Randol, *Global Climate Science: Issues for 2001*, ExxonMobil Memo to the White House, 6 February 2001. www.climatefiles.com/exxonmobil/2001-exxonmobil-randol-white-house-ipcc/

38 Julian Borger, "US and Oil Lobby Oust Climate Change Scientist," *The Guardian*, 20 April 2002.

39 The Kyoto Protocol's threshold for entry into force was ratification by 55 countries representing 55 percent of 1990 GHG emissions.

40 For example, in a 2005 speech to a Brazilian forum on climate change and biodiversity in which he praised "robust growth" and ethanol production, Wolfowitz argued that there was "much debate in the global community about the major causes of climate change and how best to curb carbon emissions." Paul Wolfowitz, *Reaching for a Double Dividend*, Forum on Global Climate Change and Biodiversity, São Paulo, Brazil, 20 December 2005.

41 Judy Pasternak, "Climate Isn't Part of World Bank Equation," *The Los Angeles Times*, 12 August 2007.

42 Krishna Guha, "Wolfowitz Deputy under Fire over Climate," *The Financial Times*, 25 April 2007.

43 Robert B. Zoellick, *An Inclusive and Sustainable Globalization*, Speech delivered at the National Press Club, Washington, DC, 10 October 2007.

44 Peter Christoff, "The Bali Roadmap: Climate Change, COP 13 and Beyond," *Environmental Politics* 17, no. 3 (2008), 466–472.

45 John Aglionby, "Bali-hoo: And So It's All Over …," *The Financial Times*, 16 December 2007.

46 Chiara Tarbacchi et al., *The Role of the Climate Investment Funds in Meeting Investment Needs*, CPI Report (London: Climate Policy Initiative, 2017). The 2008 pledges to the CIF by developed country governments totaled $6.1 billion, a significantly larger amount than GEF replenishments. The single largest contribution, $2 billion, came from the United States. See Athena Ballesteros, *Unfinished Business on Climate Change Investment Funds*, World Resources Institute Blog, 8 October 2008. www.wri.org/blog/2008/10/unfinished-business-climate-change-investment-funds.

47 Neil Bird et al., *Transformational Change in the Climate Investment Funds: A Synthesis of the Evidence*, ODI Report (London: Overseas Development Institute, 2019).

48 Author's interview with former World Bank official, 12 June 2020.

49 Pallavi Aiyar, *New Old World: An Indian Journalist Discovers the Changing Face of Europe* (New York: St. Martin's Press, 2015).
50 Lalit K. Jah, "WB Okays $180-mn Loan for India's Power Plants," *Business Standard*, 19 June 2009.
51 World Bank, *South Africa – Eskom Power Investment Support Project* (Washington, DC: World Bank, 2010).
52 World Bank, *Development and Climate Change: A Strategic Framework for the World Bank Group* (Washington, DC: World Bank, 2008).
53 World Bank, *Criteria for Screening Coal Projects under the Strategic Framework for Development and Climate Change, Operational Guidance for World Bank Group Staff* (Washington DC: World Bank, 2010).
54 Suzanne Goldenberg, "World Bank's $3.75bn Coal Plant Loan Defies Environment Criticism," *The Guardian*, 9 April 2010.
55 Lisa Friedman, "New South African Coal Plant Seeks Emission Credits for 'Cleaner' Coal," *ClimateWire*, 2 June 2010.
56 Alan Beattie and Richard Lapper, "Battle Lines Drawn as South Africa Wins Loan," *The Financial Times*, 9 April 2010.
57 Lesley Wroughton, "Global Climate Battle Plays Out in World Bank," *Reuters*, 7 March 2009.
58 Beattie and Lapper, "Battle Lines Drawn as South Africa Wins Loan."
59 Author's interview with senior World Bank official, 14 June 2018.
60 Author's interview with former World Bank official, 12 June 2020.
61 World Bank, *World Bank Appoints Andrew Steer as Special Envoy for Climate Change*, Press Release, 25 June 2010. www.worldbank.org/en/news/press-release/2010/06/25/world-bank-appoints-andrew-steer-as-special-envoy-for-climate-change.
62 CDKN, "CDKN interviews Andrew Steer, Special Envoy for Climate Change at the World Bank," *Climate and Development Knowledge Network*, 27 June 2011.
63 Author's interview with senior World Bank official, 8 July 2016.
64 World Bank, *Global Development at a Pivotal Time: A Conversation with World Bank President Dr. Jim Yong Kim*, Transcript, 19 June 2012. www.worldbank.org/en/news/speech/2012/07/18/world-bank-group-president-jim-yong-kim-brookings-institution.
65 World Bank, *Toward a Sustainable Energy Future for All: Directions for the World Bank Group's Energy Sector* (Washington, DC: World Bank, 2013).
66 Jim Yong Kim, *Foreword, Turn Down the Heat: Why a 4°C Warmer World Must Be Avoided*, Report prepared by the Potsdam Institute for Climate Impact Research and Climate Analytics on behalf of the World Bank (Washington, DC: World Bank, 2012).
67 World Bank, *Climate Change Report Warns of Dramatically Warmer World This Century*, FeatureStory, 18 November 2012. www.worldbank.org/en/news/feature/2012/11/18/Climate-change-report-warns-dramatically-warmer-world-this-century.
68 World Bank, *World Bank Group Strategy* (Washington, DC: World Bank, 2013).

69 World Bank, *Mainstreaming Climate Action within Financial Institutions: Five Voluntary Principles.* www.worldbank.org/content/dam/Worldbank/document/Climate/5Principles.pdf.

70 Sandrine Rastello, "Kim Says World Bank Staff Shakeup to Refocus on Poverty Goal," *Bloomberg*, 30 July 2013.

71 Annie M. Berglof, "World Bank Climate Change Envoy Rachel Kyte on Her New Mission," *The Financial Times*, 24 January 2014.

72 Author's interview with senior World Bank official, 20 June 2016.

73 Author's interview with former World Bank official, 12 June 2020.

74 Ed King, "UN Climate Summit Set for Major Carbon Pricing Announcement," *Climate Change News*, 12 September 2014.

75 World Bank, *Closing the $70 Billion Climate Finance Gap*, Feature Story, 9 April 2015. www.worldbank.org/en/news/feature/2015/04/09/closing-the-climate-finance-gap.

76 World Bank, *World Bank Group Exceeds Its Climate Finance Target with Record Year*, Press Release, 19 July 2018. www.worldbank.org/en/news/press-release/2018/07/19/world-bank-group-exceeds-its-climate-finance-target-with-record-year.

77 Author's interview with former World Bank official, 12 June 2020.

78 World Bank, *World Bank Group President Jim Yong Kim Remarks at Davos Press Conference*, Transcript, 23 January 2014. www.worldbank.org/en/news/speech/2014/01/23/world-bank-group-president-jim-yong-kim-remarks-at-davos-press-conference.

79 Ken Cohen, *Some Thoughts on Divestment*, ExxonMobil Blog, 10 October 2014. www.exxonmobilperspectives.com/2014/10/10/some-thoughts-on-divestment/.

80 Michele Della Vigna et al., *Re-Imagining Big Oils: How Energy Companies Can Successfully Adapt to Climate Change*, Goldman Sachs Equity Research, 8 October 2018.

81 Donna Barne, *Annual Meetings: Kim, Lagarde Talk Climate Change and Growth*, World Bank Blog, 8 October 2013. https://blogs.worldbank.org/voices/annual-meetings-kim-lagarde-talk-climate-change-and-growth.

82 World Bank, *World Bank, IMF Leaders Make Economic Case for Climate Action*, Feature Story, 9 October 2013. www.worldbank.org/en/news/feature/2013/10/08/world-bank-imf-leaders-make-economic-case-for-climate-action.

83 Valerie Volvici, "IMF, World Bank Leaders Engage Finance Ministers to Tackle Climate Change," *Reuters*, 11 April 2014.

84 World Bank, *Putting a Price on Carbon*, Statement, 3 June 2014. www.worldbank.org/en/programs/pricing-carbon#Statement.

85 World Bank, *73 Countries and Over 1,000 Businesses Speak Out in Support of a Price on Carbon*, Feature Story, 22 September 2014. www.worldbank.org/en/news/feature/2014/09/22/governments-businesses-support-carbon-pricing.

86 King, "UN Climate Summit Set for Major Carbon Pricing Announcement."

87 Author's interview with former World Bank official, 12 June 2020.

88 UNFCCC, *Paris Agreement* (Bonn, Germany: UNFCCC, 2015).

89 The nature of key terms in Article 6 has remained as a subject to intense debate and large economies of the Global South such as Brazil have pushed back against rules to avoid double counting. See, e.g., Chloé Farand, "What Is Article 6? The Issue Climate Negotiators Cannot Agree," *Climate Change News*, 2 December 2019; and Robert Stavins, "The Madrid Climate Conference's Real Failure Was Not Getting a Broad Deal on Global Carbon Markets," *The Conversation*, 18 December 2019.

90 Carol Davenport, "Carbon Pricing Becomes a Cause for the World Bank and I.M.F.," *The New York Times*, 23 April 2016.

91 Antonio Guterres, *Secretary-General's Statement on the Results of the UN Climate Change Conference COP25*, 15 December 2019. www.un.org/sg/en/content/sg/statement/2019-12-15/secretary-generals-statement-the-results-of-the-un-climate-change-conference-cop25.

92 A year later, the group published a follow-up statement, joined by the Asian Infrastructure Investment Bank, the Islamic Development Bank, and the New Development Bank. World Bank, *The MDBs' Alignment Approach to the Objectives of the Paris Agreement: Working Together to Catalyse Low-Emissions and Climate-Resilient Development* (Washington, DC: World Bank, 2016).

93 EIB, *Delivering Climate Change Action at Scale: Our Commitment to Implementation, Joint Statement by the Multilateral Development Banks at Paris, COP21* (Luxembourg: European Investment Bank, 2015). www.eib.org/attachments/press/joint-mdb-statement-climate_nov-28_final.pdf.

94 World Bank, *World Bank Group Climate Change Action Plan 2016–2020* (Washington, DC: World Bank, 2016).

95 Dave Keating, "UN Climate Summit Kicks Off with $200 Billion Pledge from World Bank," *Forbes*, 3 December 2018.

96 Heike Mainhardt, *World Bank Group Financial Flows Undermine the Paris Climate Agreement: The WBG Contributes to Higher Profit Margins for Oil, Gas, and Coal* (Berlin: Urgewald, 2019). https://urgewald.org/sites/default/files/World_Bank_Fossil_Projects_WEB.pdf.

97 Larry Elliott, "World Bank Says Paris Climate Goals at Risk from New Coal Schemes," *The Guardian*, 9 October 2016.

98 Author's interview with senior World Bank official, 14 June 2018.

99 World Bank, *World Bank Group Announcements at One Planet Summit*, Press Release, 12 December 2017. www.worldbank.org/en/news/press-release/2017/12/12/world-bank-group-announcements-at-one-planet-summit.

100 Alex Doukas, "Reaction: World Bank Steals Show at One Planet Summit by Phasing Out Upstream Oil and Gas Finance," *Oil Change International*, 12 December 2017.

101 Author's interview with former World Bank official, 12 June 2020.

102 Larry Elliott, "New World Bank Chief Confirms Commitment to Environment," *The Guardian*, 9 April 2019.

5   Conclusion
Governance architectures in transformation

Two central insights lie at the heart of the research presented in this study. First, international bureaucracies active in the climate and energy fields have been undergoing processes of governance integration, understood as the convergence of approaches and practices among different actors within one or between two or more governance architectures. Tackling the climate emergency and advancing low-carbon, sustainable development have become fundamental to energy analysis, investment, and decision-making at all levels of human organization. Due to energy's central role in creating the climate problem in the first place – two-thirds of all global GHG emissions are tied to energy production and consumption – it must also be at the heart of its solution. The IEA Secretariat's and World Bank's moves to broaden their activities and integrate the core objectives of the climate convention including, more recently, the goals of the Paris Agreement, have drawn heretofore largely separate governance architectures closer together. Indeed, the three international bureaucracies analyzed herein are more closely aligned today than at any point in their history. Second, the governance integration observed is the consequence of organizational change, understood as the change in the commitment of international bureaucracies to an objective, resulting in the adoption of new approaches and activities. The following sections draw together the key findings from the three cases, looking at organizational change through the prisms of the global policy environment, organizational autonomy, organizational structure, and organizational leadership. The remaining parts of the chapter assess the effects of governance integration in light of the multitude of cross-cutting, global challenges facing human civilization in the twenty-first century. The chapter concludes with an agenda for further research.

DOI: 10.4324/9781315661339-6

Governance integration through organizational change

Integration between key actors in the energy and climate governance architectures has come as the result of organizational change in the three cases analyzed in this book. The UNFCCC Secretariat, the IEA Secretariat, and the World Bank have moved closer together by adjusting their stances and activities over time. The IEA's narrow focus on oil supply security in the 1970s gave way to a broad-based approach to the global energy system that today puts the climate emergency and renewable energy sources front and center. Similarly, the World Bank transformed itself from a strong focus on concessional lending for fossil fuel infrastructure in developing countries and comparatively little concern for environmental impacts in its infant days into a player in full recognition of its indispensable role in driving the twenty-first century's interconnected global climate and sustainable development agendas. Importantly, both have aligned themselves with the key objectives of the climate convention and have closely cooperated with the UNFCCC. Both have also become vocal advocates of the urgent need to effectively address climate change and shift the global economy onto a low-carbon footing. The UNFCCC Secretariat's moves to broaden its approach beyond its original scientific and environmental focus and significantly increase its engagement with actors outside the traditional climate governance architecture underpinned and helped enable these developments.

The organizational changes observed have been found to be the result of the interplay of a number of factors. The first, the global policy environment, provides the basis for the other three: organizational autonomy, organizational structure, and organizational leadership. As the three cases have illustrated, the international bureaucracies' engagement with the global policy environment they operate in is a key prerequisite for organizational change and, consequently, governance integration, as it is in response to changing external circumstances and challenges that international bureaucracies come to act in the first place.

A changing global policy environment

All three international bureaucracies discussed herein have actively engaged with a complex and dynamic global policy environment. This environment is shaped by a number of aspects, including scientific evidence for the climate emergency and the risks it poses, diplomatic efforts to find collective solutions to mitigate and adapt to climate change, a host of climate and energy policies enacted in a number of different jurisdictions around the world, the changing economics of renewable energy sources, and a power shift in the global system from North to South. It has pushed the three international bureaucracies to adapt and has provided them with windows of opportunity to broaden and adjust their approaches and activities.

Greater scientific evidence for the reality of and risks inherent to anthropogenic climate change produced by the IPCC and other national and international scientific bodies gave the climate convention and its Secretariat their raison d'être. In addition to providing the scientific basis for negotiations, IPCC ARs and special reports also focused the attention of public, private, and non-governmental stakeholders alike. While the IEA Secretariat and the World Bank drew on IPCC findings, they also helped to shape scientific outputs. The IPCCs AR5 relied heavily on statistics and reports by the IEA and, to a lesser extent, the World Bank whose former Chief Scientist had, for a period, also doubled up as IPCC Chair.

Drawing on mounting scientific evidence, the international community attempted to come together in pursuit of ways to effectively address anthropogenic climate change. In the years since negotiations produced the Kyoto Protocol, climate diplomacy went through a period of ups and downs, including, most prominently, COP15 in Copenhagen which focused global attention and attempted, yet failed, to deliver a follow-on agreement. But the negotiations process also became more professional, with the UNFCCC Secretariat growing its number of staff and expertise and cooperating more closely with a range of other key stakeholders, including the IEA and the World Bank. It was, in part, thanks to this closer interaction between increasingly aligned international bureaucracies that a breakthrough was achieved at COP21 in Paris.

Following, and in some cases preceding, international efforts to address climate change, a growing number of countries enacted policies aimed at reducing GHG emissions and growing their domestic renewable energy industries. The *Global Climate Legislation Study* found that at the time of the Paris Agreement, more than 800 climate change laws and policies had been passed in 99 countries around the world, up from just over 50 at the time of the Kyoto Protocol in 1997.[1] At the end of 2018, 169 countries also had some kind of renewable energy (promotion) policy in place, up from 48 countries in 2005 and less than half that number in the late 1990s.[2] These developments fundamentally transformed the economics of renewable energy by creating a guaranteed market demand, mobilizing investment for industrial-scale capacity additions, and enabling wind and solar PV to compete with established fossil fuels and nuclear power.

The changing economics of renewables like wind and solar PV brought the viability of a low-carbon transition necessary to effectively mitigate anthropogenic climate change into sharper focus. As the share of renewables in power sectors across the world has grown and private investors have increasingly chosen them over fossil fuel alternatives, international bureaucracies operating in the climate and energy fields have adjusted their approach. The IEA Secretariat now prominently addresses renewables as part of its core portfolio, a process sped up considerably by the creation of IRENA. IEA renewable technology outlooks are also more bullish than in the past, owing to the growing cost competitiveness of wind and solar PV. The changing economic case for renewables has had significant implications for the UNFCCC Secretariat, too, with Executive Secretary Figueres regularly using her bully pulpit to position the downsides of a continued reliance on fossil fuels against the opportunities of investing in renewable technologies. Finally, the World Bank's support for renewables shifted along with growing policy support in its contributor states and declining technology costs, with the Bank eventually moving away from coal and upstream oil and gas and rebalancing its energy lending portfolio toward low-carbon, renewable sources.

In the nearly 30 years since the UNCED, the global political environment has changed dramatically. At the heart of this change has been a gradual shift in economic might and political influence from the United States and its allies in the Global North toward China, India, and other rising powers of the Global South. A period of unipolar strength projected by the United States after the end of the Cold War has increasingly given way to a world in which no country alone can shape global events and where the various poles compete with each other more intently than before. Bremmer refers to this as the *G-Zero* world, a world in which no one country is able or willing to effectively assume the mantle of global leadership.[3] A G-Zero world with its leadership vacuum and often conflicting state interests undermines precisely the kind of international collective action needed to tackle the climate emergency and address its root causes in the energy system in both the Global North and South. US disengagement from the Kyoto Protocol and the Paris Agreement under the Bush and Trump administrations, respectively, is a symptom of these developments. However, while trying to make sense of a cacophony of voices and bridge goal conflicts between multiple state principals can make life difficult for international bureaucracies, it also has, as this study has shown, equipped them with opportunities to take on new responsibilities and chart a more autonomous path to a low-carbon future.

Organizational autonomy

In contrast to earlier research on the role of IOs and their bureaucratic arms that viewed them as enjoying little independence from their state principals, more recent analysis has pointed to an understanding of such agents as more autonomous and independent actors, especially in situations where there are multiple principals and persistent information asymmetries. All three international bureaucracies analyzed in this study have a group of states as their collective principal. In the case of the UNFCCC, member countries are represented as parties to the climate convention. IEA member countries exercise control via the Governing Board made up of energy ministers. And member countries influence the direction of the World Bank through the Board of Governors, the Board of Executive Directors, and the President of the Bank who has traditionally been nominated by the United States as the largest shareholder. Policy preferences are multifarious in each case, with regard to the UNFCCC and the World Bank more so than the IEA given its smaller membership made up of OECD countries only. Despite the climate convention having a larger membership and, thus, more principals than the other IOs, the UNFCCC Secretariat's narrower and more specific mandate does not, it would seem, provide it with a lot of decision-making autonomy. The World Bank's broad mandate on the other hand has given it much greater independence and authority vis-à-vis its collective principal. Moreover, the UNFCCC and IEA Secretariats have continued to rely on member state contributions as their main source of income while the influence of member states over the Bank's operations has lessened over time as their payments have declined in relative terms compared to funds raised by the Bank in the global financial markets.

However, while member states exert a degree of control, including through the power of the purse, the observed activities of the UNFCCC Secretariat, the IEA Secretariat, and the World Bank are not merely the responses of agents to principals' directives. As this study shows, international bureaucracies in the climate and energy fields have engaged with their mandate in unexpected ways, in each case carving out greater autonomy over time. The UNFCCC Secretariat began its work in the early 1990s without a noticeable profile and with no agenda of its own, largely content with the background role of quietly supporting country delegations to the climate convention. Along with a rise in awareness, underpinned by scientific findings for the reality and impacts of anthropogenic climate change and a greater sense of urgency to provide adequate solutions to the problem, the Secretariat grew stronger and bolder, taking on more responsibilities and cooperating with an increasingly diverse group of stakeholders. By the time of the Paris Agreement in 2015, it had developed into a potent "orchestrator," guiding negotiators and more self-assuredly pushing the climate convention process forward.

The IEA Secretariat was formed without a clear role to influence or shape member state energy policies beyond providing collective insurance in case of oil supply disruptions. Member states enabled the IEA to broaden its research programs and occasionally chose direct confrontation to goad it into doing more and better, as in the case of the creation of IRENA, but it is the specific ways in which the Secretariat responded to these opportunities and challenges that transformed it into an international bureaucracy widely considered as the most authoritative global voice on issues affecting the entire energy system. No other comparable organization can claim to command the same wealth of information and have as much impact, not just on its members but also on its partner countries in the Global South and on a wide range of stakeholders across the public, private, and not-for-profit sectors. The IEA's senior leadership has further used its editorial autonomy over IEA publications to focus WEOs and a range of special reports on matters of climate change and low-carbon transition scenarios. Among the occasional differences between member states on issues such as the Kyoto Protocol or the Paris Agreement, the Secretariat negotiated its own path, cautiously at first and more vigorously later, to arrive at a position of independent leadership within the global climate debate.

The World Bank evolved in similar ways. In contrast to the emissions-intensive, fossil-fueled focus of its early days, the Bank moved to occupy a central role in global climate and energy governance. Along this road spanning three decades, it faced strong calls from environmental NGOs and some of its major shareholders, who had themselves embarked on a fundamental, policy-induced transformation of their energy systems, to shift more aggressively to sustainable, low-carbon lending and cease any remaining financial support for coal projects. In its attempts to muster an effective response and balance the impacts of climate change on its core mission of poverty alleviation with the continued demands of its developing member countries, the international bureaucracy developed innovative funding solutions like the CIF, pushed back against the occasional climate skepticism of the United States as its largest shareholder, and eventually morphed into a forceful advocate for greater global ambition in climate change mitigation and adaptation.

Taken together, the three international bureaucracies have played an increasingly autonomous and consequential role in global climate and energy governance through their work in assisting negotiations under the climate convention, providing authoritative data and analysis, identifying, supporting and, in the case of the World Bank, financing solutions, building coalitions with key stakeholders, and behaving like "activist bureaucracies" in calling for greater urgency and helping to build momentum for change through sustained pro-climate advocacy. A closer look at internal structural changes in each of the three cases helps to explain these developments further.

Organizational structure

Changes to the structural features of international bureaucracies are closely entwined with moves toward greater organizational autonomy as changing structures represent a conduit for autonomous bureaucratic preferences to be turned into action.[4] The UNFCCC Secretariat, the IEA Secretariat, and the World Bank each expanded their work and increased the number of expert staff working on climate change and low-carbon energy technologies over time. The UNFCCC Secretariat expanded from a more randomly assembled handful of people fitting around a dinner table in the early 1990s to a staff of more than 400 experts organized into clear workstreams and programs at the time of the Paris Agreement. The Secretariat also developed and led new initiatives such as the NAZCA and the externally funded Momentum for Change Initiative. The substantial growth and professionalization of its staff enabled it to occupy a position of greater authority in the climate negotiations, build more links with other stakeholders contributing to processes of governance integration, and help shape outcomes in line with the 2°C climate stabilization goal.

The IEA's original structure reflected its genesis as an international bureaucracy focused on oil supply security specifically, and fossil fuel energy sources more generally. This structure underwent adjustment following the broadening of the IEA's mission in the 1980s and 1990s and then again in the aftermath of the G8 summit in Gleneagles. The IEA established the CCXG to support negotiators to the climate convention, created new units and grew their staff, including those on renewable energy sources, energy efficiency, and environment and climate change. The growth and shift in focus of its expert publications, including the flagship WEOs, its policy advisory, its expanding interaction with China, India, and other major non-OECD economies, in addition to the manifold contributions to the work of the UNFCCC, including through the gathering of GHG emissions statistics, are a testament to these changes. It is hard to see how without them the IEA could have retained its role as the world's leading organization in the energy space. The World Bank's structural changes in pursuit of integrating climate issues into its portfolio are arguably the farthest reaching of the three cases. This may seem surprising considering the multitude of developmental challenges the Bank has been addressing in its more than 70 years of existence. However, over time, experts within the Bank came to realize not just the potential for expanding the Bank's work into climate and carbon finance but also the risks unmitigated climate change poses to the achievement of its twin objectives of ending extreme poverty and promoting shared prosperity in a sustainable manner. The number of staff working on environmental and climate issues grew and from the early 1990s onward, the Bank established, capitalized, and expanded a number of innovative financial instruments such as the GEF, the PCF, and the CIF. It created dedicated units such as the CFU and climate-focused roles in senior management, including that of Special Envoy for Climate Change at the level of World Bank Vice President. In 2013, the Bank underwent a structural reform process that aligned the WBG's five branches for more effective operation and created a new system of so-called Global Practices strongly focused on sustainable development and prominently positioning climate change as one of five cross-cutting themes. In so doing, the Bank engaged in a systematic and explicit integration of climate change considerations into policy planning, investment design, implementation, and evaluation; that is, it mainstreamed climate change into its operations. While not always top-down processes, changes in the Bank, as much as in the UNFCCC and IEA Secretariats, were directly connected to, and in many instances dependent upon, the positions taken by the international bureaucracies' senior leadership.

Organizational leadership

This study has found leadership to be one of the key determinants for organizational change and governance integration. Leaders translate a multitude of challenges and opportunities arising in a complex and dynamic global policy environment, both within the bureaucracy, identifying the need for innovation and driving changes in organizational structure and focus, and in its outward engagement, positioning the international bureaucracy with regard to key issues, and focusing events, and forging closer links with a variety of stakeholders and interested parties. The tracing of the entrepreneurial activity displayed by the senior leadership in each of the three cases adds important insights to a literature that has not devoted sustained attention to the specific effects of leadership on organizational change.

The UNFCCC Secretariat commands very little direct power, but it affords its Executive Secretaries with the kind of bully pulpit that de Boer and, even more so, Figueres used to maximum effect. Drawing on a bureaucracy made more potent thanks to the efforts of the more cautious Cutajar, they seized the opportunity to expand the reach and meaning of the climate convention beyond its original scientific and environmental focus, entrepreneurially framing the debate on the climate problem and its solution in language understood by key players, and building the kind of rhetorical and physical linkages between climate change and related governance architectures that would later contribute to securing a breakthrough at COP21. The UNFCCC Executive Secretaries transformed an understaffed technocratic bureaucracy responsible for merely facilitating climate negotiations and barred from having opinions of its own into a publicly visible, potent force, confidently making the case for a rapid decarbonization of the global economy and guiding country delegations toward reaching collective solutions. Yet such solutions still need to be translated into ambitious and effectively implemented climate and energy policies at national and subnational levels.

The IEA Secretariat's push for just such policies represents a shift from its earlier role which was marked by a considerable degree of skepticism toward the role renewables could play in the twenty-first century. Earlier Executive Directors recognized climate change as an issue and helped broaden the Agency's portfolio but were more cautious in their approach to key outcomes of the climate convention process. More recently, van der Hoeven and Birol framed climate change mitigation and the transition to low-carbon energy sources as an urgent and necessary challenge, advocating more forcefully on behalf of these interconnected agendas and increasingly aligning the IEA Secretariat with the core objectives of the climate convention. In their response to the outcomes of climate negotiations, stark IPCC warnings of the dangerous impacts of a warming world, changes in the economics of wind and solar PV, greater climate and energy policy ambition in IEA member states, and the challenge posed by the creation of IRENA, they ensured a continued role for the Agency as the world's leading multilateral energy organization.

Similarly, the World Bank's senior leadership drove the organizational change that enabled the international bureaucracy to adapt to a fast-changing global policy environment and retain its role as the leading MDB operating in the Global South. While earlier World Bank Presidents recognized climate change as an issue that needed to be addressed, dissonances remained. It was under Kim that climate change took center stage, following several years of building internal capacity and developing innovative financial solutions to support a low-carbon transition. Kim and his core leadership team used their platform to help drive the climate debate, directly and actively support the climate convention process, include climate considerations across the Bank's operations, end the direct lending to coal and upstream oil and gas projects campaigners had long called for, and set ambitious targets to increase the share of climate-related lending to nearly a third of the total. These and other efforts resulted in greater alignment with the goals of the Paris Agreement and contributed to processes of governance integration.

Effects of governance integration

The number of cross-cutting policy and governance challenges facing decision-makers in the twenty-first century can seem daunting. Ending poverty and hunger, providing clean drinking water for all, halting the spread of dangerous communicable diseases and pandemics such as Covid-19, ending conflict and violence, responding to climate change, achieving gender equality – these and other objectives incorporated in the UN SDGs cannot be addressed by any one country or IO alone. They require stakeholders at all levels of political authority and human organization across the Global South and North to pull together in the pursuit of effective collective action. Climate change has, at times, been described as a "super wicked problem,"[5] the "toughest, most intractable political issue we, as a society, have ever faced,"[6] and the "defining issue of our time."[7] Climate change is such a complex and multifaceted issue, in fact, that successfully addressing it requires a mainstreaming of climate change considerations into all related policy areas. In the first instance, climate change has rightly been integrated into energy policy given that energy production and consumption are both the main cause for and solution to the climate emergency. Following the same logic, the global climate and energy governance architectures have become more integrated as international bureaucracies such as the IEA Secretariat and the World Bank have aligned themselves with the objectives of the climate convention and cooperated more closely with the UNFCCC Secretariat.

This study points to a number of effects of governance integration. First, governance integration changes the trajectory of international bureaucracies and the IOs they serve. Each of the three cases analyzed began their life focused on particular activities that today make up only a part of their operations. The integration of the key goals of climate change mitigation and adaptation, and the means to achieve them, broadened the field of play for both the IEA and the World Bank. It opened up new avenues, such as the IEA's work on renewables and GHG emissions statistics, and closed others, such as the World Bank's funding for coal-fired power plants and upstream oil and gas projects. Broadening the scientific and environmental focus of the climate convention and cooperating more closely with actors in related governance architectures enabled the UNFCCC Secretariat to occupy a considerably more central role in the drive for a post-Kyoto climate agreement than may once have been considered possible given its humble beginnings. The growth in the number of expert staff in each of the three bureaucracies, addressing GHG emissions, carbon finance, renewables, and a range of other issues, has professionalized the bureaucracies further and has heightened the impact of their actions. They now competently and confidently address climate change and energy as interconnected challenges.

Second, governance integration creates greater synergies within and between related governance architectures. While occupying central roles in their respective governance fields, the UNFCCC, IEA, and World Bank are but a part of a dense web of IOs and their bureaucracies, as well as other public, private, and non-governmental actors operating in domestic, international, and transnational spaces in the pursuit of interlinked initiatives and goals. This includes other parts of the UN family such as UNEP and UNDP, other MDBs such as the EIB, ADB, AIIB, and AfDB, but it also includes other international energy organizations such as IRENA and OPEC, the latter of which has in recent years begun to address climate change issues and engage more actively with the climate convention, albeit in a significantly more limited fashion than the IEA owing to its exclusive focus on crude oil and petroleum products. Both the climate and energy governance architectures are, thus, complex networks of multiple governing authorities at different scales. This polycentricity offers opportunities for learning and innovation within international bureaucracies and forges paths toward greater integration and mutual reinforcement between global climate and energy governance as seen in the changing approaches and activities of the UNFCCC Secretariat, the IEA Secretariat, and the World Bank as independent yet increasingly interconnected actors.

Third, changing the climate debate and creating greater salience for moves to address the climate emergency and swift transition to a low-carbon future is a further key effect of governance integration. The UNFCCC Secretariat's, IEA Secretariat's, and World Bank's arguments in support of disruptive policy change and the sustained framing of urgently tackling anthropogenic climate change have had a cognitive impact on numerous other actors by reducing ambiguity over core concepts and particular policy solutions, focusing attention and keeping the issue on the agenda. Arguments raised by the World Bank's and the IEA's leadership have increasingly mirrored those made by the UNFCCC Secretariat and leading actors in the global climate negotiations. Research on the impact of message framing conducted in a variety of fields has long recognized that source credibility makes a difference to an actor's willingness to accept a product or argument.[8] Both the high organizational credibility of the three international bureaucracies in the climate and energy fields, respectively, and the authority of those senior officials entrepreneurially engaging in the advocacy, including Figueres, Kim, and Birol, among others, have meant that the message is more likely to be accepted by target audiences than if it had come from other sources.

Fourth, and relatedly, governance integration can contribute more directly to solving the climate emergency through improving climate and energy policies and mobilizing greater investment. The IPCC's AR6 Working Group 1 report has made clear that climate change is speeding up and intensifying.[9] According to the UNEP Emissions Gap Report 2020, the world is heading for a more than 3°C rise in global average surface temperatures above pre-industrial levels by the end of the century.[10] The temporary reduction in global GHG emissions due to the depressive effects of the Covid-19 pandemic on economic activity around the world notwithstanding, current emissions trajectories mean that the goals of the Paris Agreement are out of reach, even if countries were to meet all the climate commitments they have made so far. This further increases the risks of harmful climate impacts, many of which are already materializing today.

If a three- or four-degree world in all its dire consequences is to be avoided, the transition to a zero-carbon future needs to be urgently sped up. In this, international bureaucracies like the IEA Secretariat and the World Bank have an important role to play beyond supporting the climate convention and framing the case for change. The IEA's statistics, expert reports, and policy advice enable public and private stakeholders to make policy and investment choices and take the necessary steps to decarbonize the energy system. The shift in the IEA's treatment of renewable energy sources has been critical in this regard. The World Bank plays a similar role, yet, unlike the IEA, it also provides concessional loan and grant funding, crowding in much-needed private sector investment by drawing on its power of leverage. Climate and carbon finance were only a small part of the Bank's operations in the 1990s. Today, roughly a third of all the Bank's investments are climate related and climate impacts are evaluated across the portfolio. Together with the investment mobilized by other MDBs, this has made a real difference on the ground. However, neither the IEA nor the World Bank were created to serve environmental purposes. Both continue to also pursue other goals, the IEA Secretariat through its work on fossil fuels and the World Bank in its funding of non-energy infrastructure projects that nevertheless come with a carbon footprint. Successfully turning the tide on dangerous climate change, therefore, requires governance and policy integration beyond the three cases analyzed in this study.

Finally, the governance integration observed in this study has contributed to preserving and expanding the role and relevance of the international bureaucracies analyzed. Climate change is such an enormous and all-enveloping global issue that IOs and their bureaucracies ignore it at their peril. Neither the IEA nor the World Bank was founded to pursue environmental goals, and both have come in for their fair share of criticism for not responding to the climate challenge fast enough, be it through a downplaying of the growth potential of renewable energy sources or the continued investment in coal-fired power generation. Yet, given the interconnections between climate change on the one hand and energy, economic development, and poverty alleviation on the other, the climate emergency now occupies a central position in the approaches, understandings, and activities of both IOs. The IEA Secretariat's and World Bank's drive for change is, therefore, as much a response to the global policy environment discussed above as a self-interested push for continued relevance and survival as key global actors. What good is an International Energy Agency if it does not address the climate emergency as the defining energy problem of the twenty-first century, does not see renewable energy sources as the key solution, and does not understand the transformative nature of the low-carbon transition across the Global North and South? What good is a World Bank as the largest MDB operating outside the OECD world if it does not understand the extraordinary risks climate change poses to the achievement of the SDGs, does not incorporate climate at the heart of its operations, and does not screen and adjust all of its projects accordingly? Both organizations would be abdicating themselves and would lose further ground to others, the IEA to IRENA and other organizations and the World Bank to public and private IFIs. In their alignment with the objectives of the Paris Agreement, their cooperation with the UNFCCC Secretariat, and their increasingly resolute change advocacy, they have chosen adaptation over stagnation and continued significance over managed decline. This has strengthened both global energy and climate governance.

Conclusion

The climate emergency can only be effectively addressed if actors at all levels, from the local to the global, across the Global North and South and spanning the public and private divide, accept their shared responsibility and together take ambitious and meaningful action to mitigate the core problem – GHG emissions largely created in the production and consumption of energy – and adapt to its many effects already materializing today. Getting it right will have bearing not only on global average surface temperatures, sea level rise, precipitation patterns, the frequency and severity of extreme weather events, and a whole host of other environmental impacts but also on the world's ability to meet the SDGs, given the goals' strong focus on climate action and energy transitions and the clear and present dangers unsuccessful climate mitigation and adaptation pose to effective poverty alleviation, equitable growth and shared prosperity, health and well-being, and peaceful and inclusive societies.

The IPCC's AR6 is a stark reminder that the world has not yet risen to the climate challenge. So far, most NDCs submitted under the Paris Agreement are not fully compatible with a 2°C pathway, let alone with the 1.5°C ambition spelled out in its Article 2. Deciding the Paris rulebook has proved tricky, too, with agreement not forthcoming until COP24 in Katowice and the operationalization of Article 6 on carbon markets delayed until 2021. The convention process has slowed down further with Covid-19 delaying COP26 and capturing policymakers' attention despite calls for a low-carbon, climate-friendly recovery supported by a wide range of actors at all levels and scales, including the three international bureaucracies analyzed in this study.

Despite these difficulties, however, global governance remains critically important in providing a framework for parties and non-party stakeholders to take action. For all its weaknesses, the Paris Agreement lays the foundation upon which effective governance can be built, guiding and supporting countries committing themselves to low-carbon policies as much as businesses looking for certainty in a global marketplace. The complexity of the climate challenge and the fragmented, polycentric nature of both global climate and global energy governance necessitate more cooperation and coordination between actors addressing different pieces of the bigger puzzle, not less.

IOs and their bureaucracies operating in the climate and energy fields have a varied role to play, from the provision of data and analysis, policy advice, global advocacy, financial support and catalyzation of private sector capital, to ending the compartmentalization that has characterized the relationship between climate, energy, and related sectors for too long. The climate emergency cannot be effectively addressed without a focus on energy production and consumption at the heart of the problem. Likewise, the global energy system is in a period of deep transformation due to the need to realize a climate-neutral economy with net-zero CO_2 emissions by the middle of the century.

As this study has shown, by integrating and more closely aligning their approaches and activities, the UNFCCC Secretariat, IEA Secretariat, and World Bank are enhancing the effectiveness of both climate and energy governance and are, thus, contributing to achieving the goals of the Paris Agreement and, relatedly, the SDGs. Although connected to governance fragmentation, governance integration is not simply the opposite of the former as fragmentation and integration processes may occur in parallel. It is also closely related to policy integration but rather than following an explicit plan designed and imposed by states, governance integration has emerged in unexpected ways, through international bureaucracies' various efforts to pursue and broaden their mandate in a complex and fast-changing global policy environment. Further research into changing, polycentric governance architectures would benefit from studying governance integration as a discrete and novel concept with an independent quality.

Even though this study is concerned with international bureaucracies occupying central positions within global climate and energy governance, its methodology and many of its findings, driven by a greater need to disaggregate the organizational black box and trace routine connections within and beyond, may be generalized to bureaucracies playing a role in other governance areas. While such areas, including trade and health, are often less fragmented, and may therefore present fewer needs and opportunities for governance integration, organizational changes shaped by engagement with a complex, dynamic global policy environment and changes in organizational autonomy, structure, and leadership are nevertheless commonplace. In-depth research into international bureaucracies in other areas may therefore produce conceptually similar findings, explaining organizational change as a result of the interplay of the four factors, while adding important empirical evidence unique to each case. Such evidence can, in turn, play an important role in the further development and improvement of theory, benefitting international bureaucracy and IR scholarship more widely.

Yet it also holds important lessons for IOs and their bureaucracies themselves. Successfully navigating the challenges of the twenty-first century and remaining relevant players in an age of climate change and other interconnected planetary goals require international bureaucracies beyond the ones considered here to adapt and, if necessary, transform, resulting in the kind of integrated and synergistic governance needed to build a more sustainable, better future for all.

Notes

1 Michal Nachmany et al., *The 2015 Global Climate Legislation Study: A Review of Climate Change Legislation in 99 Countries: Summary for Policymakers*, Grantham Research Institute on Climate Change and the Environment (London: GLOBE International, 2015).

2 REN21, *Renewables 2019 Global Status Report* (Paris: REN21 Secretariat, 2019).

3 Ian Bremmer, *Every Nation for Itself: Winners and Losers in a G-Zero World* (London: Penguin, 2013).

4 Michael W. Bauer and Jörn Ege, "Bureaucratic Autonomy of International Organizations' Secretariats," *Journal of European Public Policy* 23, no. 7 (2016), 1019–1037.

5 Kelly Levin et al., "Overcoming the Tragedy of Super Wicked Problems: Constraining Our Future Selves to Ameliorate Global Climate Change," *Policy Sciences* 45 (2012), 123–152.

6 Elaine Kamarck, "The Challenging Politics of Climate Change," Brookings Institution report, 23 September 2019. www.brookings.edu/research/the-challenging-politics-of-climate-change/.

7 Statement by UN Secretary General António Guterres on Climate Change and His Vision for the 2019 Climate Change Summit, 10 September 2018. www.un.org/sg/en/content/sg/statement/2018-09-10/secretary-generals-remarks-climate-change-delivered.

8 See, e.g., Robert D. Benford and David A. Snow, "Framing Processes and Social Movements: An Overview and Assessment," *Annual Review of Sociology* 26 (2000), 11–39; Dhruv Grewal et al., "The Moderating Effects of Message Framing and Source Credibility on the Price-perceived Risk Relationship," *Journal of Consumer Research* 21, no. 1 (1994), 145–153; and John F. Mahon and Steven L. Wartick, "Dealing with Stakeholders: How Reputation, Credibility and Framing Influence the Game," *Corporate Reputation Review* 6 (2003), 19–35.

9 IPCC, *Climate Change 2021: The Physical Science Basis. Contribution of Working Group I to the Sixth Assessment Report of the Intergovernmental Panel on Climate Change* (Cambridge: Cambridge University Press, 2021).

10 UNEP, Emissions Gap Report 2020 (Nairobi: UNFCCC, 2020).

Selected bibliography

- Andrew Jordan et al., eds., *Governing Climate Change. Polycentricity in Action?* (Cambridge: Cambridge University Press, 2018).
  The contributions in this edited volume fruitfully expand on Ostrom's work on pol ycentric climate governance, drawing on a wide range of cases to illustrate polycentricity in action and critically discussing its opportunities and limitations.
- Bob Reinalda and Bertjan Verbeek, eds., *Decision Making within International Organizations* (London: Routledge, 2004).
  The contributors to this edited volume trace and analyze decision-making processes within a number of international organizations, drawing on different theories and approaches in assessing how the EU, UN, and others engage with their organizational autonomy.
- Devesh Kapur et al., eds., *The World Bank: Its First Half Century, Volumes I and II* (Washington, DC: Brookings Institution Press, 1997).
  Although now more than 20 years old, this edited volume still represents the most encompassing account of the evolution of the World Bank, its approaches, successes, and failures during the first five decades of the Bank's existence.
- Frank Biermann and Bernd Siebenhüner, eds., *Managers of Global Change: The Influence of International Environmental Bureaucracies* (Cambridge: MIT Press, 2009).
  This edited volume develops a conceptual framework for understanding international bureaucracies and their role in environmental governance. It applies this framework to an analysis of the influence of nine different international bureaucracies, including the World Bank and the UNFCCC Secretariat.
- Frank Biermann et al., "The Fragmentation of Global Governance Architectures: A Framework for Analysis," *Global Environmental Politics* 9, no. 4 (2009), 14–40.
  The authors conceptualize governance fragmentation and develop an analytical framework to understand fragmented global governance architectures.

Fragmentation represents one of the main starting points for understanding governance integration.

- Frank Biermann and Rakhyun E. Kim, eds., *Architectures of Earth System Governance: Institutional Complexity and Structural Transformation* (Cambridge: Cambridge University Press, 2020).
The contributors to this authoritative edited volume discuss institutions, bureaucracies, and networks as the key building blocks of earth system governance, highlight key structural features of the governance system, including fragmentation and interlinkages, and point to opportunities for policy reform and better interplay management.
- Harald Heubaum and Frank Biermann, "Integrating Global Energy and Climate Governance: The Changing Role of the International Energy Agency," *Energy Policy* 87 (2015), 229–239.
The authors develop governance integration as a discrete concept, showing how changing activities and approaches by the IEA are drawing the global energy and climate governance architectures closer together.
- IPCC, *Global Warming of 1.5°C. An IPCC Special Report on the Impacts of Global Warming of 1.5°C above Pre-industrial Levels and Related Global Greenhouse Gas Emission Pathways, in the Context of Strengthening the Global Response to the Threat of Climate Change, Sustainable Development, and Efforts to Eradicate Poverty* (Geneva: IPCC, 2018).
This is, arguably, the most important scientific report published by the IPCC since AR5 in 2014. It issued a stark warning to the international policy-making community and helped drive climate negotiations and processes of governance integration discussed in this study.
- Michael W. Bauer and Jörn Ege, "Bureaucratic Autonomy of International Organizations' Secretariats," *Journal of European Public Policy* 23, no. 7 (2016), 1019–1037.
The authors develop a theoretical conceptualization of the bureaucratic autonomy of international secretariats as the administrative arms of international organizations. Their work reveals variation in the autonomous behavior of 15 international secretariats.
- Thijs van de Graaf, *The Politics and Institutions of Global Energy Governance* (London: Palgrave Macmillan, 2013).
The author offers an authoritative take on the architecture of global energy governance, analyzing the international regime complex for energy and the role of key organizations such as the IEA, IRENA, and the G20.
- UNFCCC, *Paris Agreement* (Bonn, Germany: UNFCCC, 2015).
The world's first universal climate change agreement sets the long-term goal of keeping the increase in global average surface temperatures to well below 2°C above pre-industrial levels and prescribes a path to achieve the urgently needed decarbonization of the global economy.
- Fariborz Zelli et al., eds., *Governing the Climate-Energy Nexus: Institutional Complexity and Its Challenges to Effectiveness and Legitimacy* (Cambridge: Cambridge University Press, 2020).

This edited volume analyzes the climate and energy nexus through three interconnected lenses: renewable energy, fossil fuel subsidy reform, and carbon pricing. It assesses each with a view to their management, coherence, legitimacy, and effectiveness and provides policy recommendations for more effective governance.

Index

of 110–111, 121n89, 135; and
governance architectures 122, 124,
125, 127, 128, 131, 133, 134–135,
136; and IEA 82, 85, 88; and
UNFCCC 61, 62–64, 65, 66n2; and
World Bank 94, 109, 110, 111–112,
113, 114, 115, 116
PCF (Prototype Carbon Fund) 100,
101, 103–104, 114, 115, 129
Plan of Action on Climate Change,
Clean Energy, and Sustainable
Development 55, 76, 86
policy, public 5, 7–8, 9, 12, 13, 22, 28,
31, 32–33, 39
policy advice 12, 42, 128–129, 133, 136
policy change 7–8, 12, 28, 32, 39, 97,
133
policy entrepreneurs 7, 9, 12, 32,
39–40
policy integration 21, 24–25, 134, 136
policy networks 32–33
policy windows 32
policymaking 3, 5, 24, 25, 31, 33, 37,
62, 71, 83; processes of 25, 31,
32–33, 34, 40, 49
polycentricity 4, 21, 22, 23–24, 132,
136
poverty 14, 94, 99, 105, 107, 108, 115,
117n23, 129; alleviation of 1, 101,
114, 116n7, 127, 131, 134, 135;
energy 2
Priddle, Robert 74, 76
principal-agent theory 36
process-tracing, within case studies
9–11
Prototype Carbon Fund (PCF) 100,
101, 103–104, 114, 115, 129
public policy 5, 7–8, 9, 12, 13, 22, 28,
31, 32–33, 39

regime complex 22, 23–24
renewable energy 34, 77, 78, 79;
capacity 2, 75; policy 11, 28, 52,
124; projects 84, 100, 106, 112;
sources 29, 70, 74, 75, 78, 79, 81,
84, 86, 88, 123, 124, 128, 133, 134;
see also International Renewable
Energy Agency (IRENA)
Renewable Energy Industry Advisory
Board 78

renewables 31, 36, 125, 130, 132;
and IEA 74, 79, 82, 84, 85, 86,
87; modern 70–71, 77, 96; and
UNFCCC 55, 59, 61, 62, 63; and
World Bank 94, 101, 109, 112, 114;
see also solar photovoltaic energy;
wind energy
resources, energy 1, 3, 4, 30, 31–32,
63, 72

SCF (Strategic Climate Fund) 103
scientific evidence 31, 33–34, 35, 49,
83–84, 123–124
SDGs *see* Sustainable Development
Goals (SDGs)
secretariats: international 26, 38, 40–41
Security Council, of United Nations
57
SEforALL (Sustainable Energy for
All) 106, 107
senior leadership 11, 80, 88, 115, 127,
129–130, 130–131
Seventh Conference of the Parties
(COP7) 74
SFDCC (Strategic Framework for
Development and Climate Change)
104, 105
solar photovoltaic energy 28, 55, 70,
75, 77, 78, 79, 85, 88, 104–105,
124–125, 130
SR1.5, of IPCC 2, 135
state principals 38, 125, 126
Steeg, Helga 73
Steer, Andrew 100–101, 106
Stern, Nicholas 102
Stern, Richard 56, 100–101
*Stern Review on the Economics of
Climate Change* 56, 102
Stockholm Conference 4, 96
Strategic Climate Fund (SCF) 103
Strategic Framework for
Development and Climate Change
(SFDCC) 104, 105
Strong, Maurice 50, 96
structural changes 9, 29, 38, 70, 87,
95, 128, 129
structure, organizational *see*
organizational structure
Sustainable Development Goals
(SDGs) 1, 13, 14–15, 131, 134,